Clean Android Architecture

Take a layered approach to writing clean, testable, and decoupled Android applications

Alexandru Dumbravan

BIRMINGHAM—MUMBAI

Clean Android Architecture

Group Product Manager: Rohit Rajkumar
Publishing Product Manager: Nitin Nainani
Senior Editor: Aamir Ahmed
Content Development Editor: Feza Shaikh
Technical Editor: Simran Udasi
Copy Editor: Safis Editing
Project Coordinator: Manthan Patel
Proofreader: Safis Editing
Indexer: Rekha Nair
Production Designer: Shyam Sundar Korumilli
Marketing Coordinator: Teny Thomas

First published: June 2022

Production reference: 2280922

Published by Packt Publishing Ltd.
Livery Place
35 Livery Street
Birmingham
B3 2PB, UK.

ISBN 978-1-80323-458-8

www.packt.com

This book is dedicated to all the developers who went the extra mile to learn, practice, hack, innovate, and then pass their knowledge on to others. This propelled our development world forward and made this book a possibility. This book is also dedicated to all the people in my life who supported me through the times I was staring desperately at a computer screen filled with compile errors.

Alexandru Dumbravan

Foreword

"If you think good architecture is expensive, try bad architecture."

— Brian Foote, CEO of HUMBL

We know deep down that this statement is true. Poorly architected code can cause problems in your application, but it can also make your code hard to understand, navigate, be improved upon, and difficult to use for other developers.

"The glass is neither half empty nor half full. It's simply larger than it needs to be."

— Grace Hopper

Grace Hopper was an American computer scientist and Navy officer. She knew the importance of a well-structured and clean architecture. They're flexible, higher performing, far more scalable, easier to test, and you can maintain them a lot faster and more easily. A clean architecture is focused on getting the application layers built as efficiently as possible, without leaving behind (as Grace might say) leftover cup.

In a mobile application, clean architecture is more important than ever. You might find yourself battling against battery issues, memory consumption issues, security problems, compatibility, or environmental changes. Your mobile application needs to truly achieve portability.

In *Clean Android Architecture, Alexandru Dumbravan* applies his experience of developing Android applications for over 10 years. He takes you through a quick tour of the core software design principles and through the key features in the Android framework and supporting libraries.

Next, you'll dig deep in data sources. It's vital to understand the libraries and frameworks that are available for you to access and manage data. Then, Alexandru naturally transitions into data presentation. How do you present data to the user? You'll break down how the user interface (UI) works and how to build your own UI solution.

Later, you'll learn how to manage dependencies. Part of the value that Alexandru brings to the book is that he shows you how these different challenges were handled in the past, and what has changed over time. You'll then see how to build the domain layer to sit in the center of your application and to control your app logic. Next, you'll build the data layer to create and manage your data. The presentation layer handles the UI and the input/output. Finally, you'll put the modules together and test your application.

By building an Android application with a clean architecture, you'll be able to put every piece of your code to work, and you'll end up with a far more scalable, performant, and maintainable application. This is a book that you'll want to come back to, to make sure you've got each step of this process down. And once you do, well, we all know that the present and future is in mobile applications. So, the question is, where are you taking us?

Ed Price

Senior Program Manager of Architectural Publishing

Microsoft, Azure Architecture Center (`http://aka.ms/Architecture`)

Co-Author of 7 Books, including Meg the Mechanical Engineer, The Azure Cloud Native Architecture Mapbook (Packt), and ASP.NET Core 5 for Beginners (Packt)

Contributors

About the author

Alexandru Dumbravan has been an Android developer since 2011 and has worked across a variety of Android applications that have contained features such as messaging, voice calls, file management, and file management. He continues to broaden his development skills while working in London for a popular fintech company.

About the reviewers

Revathi Gopalakrishnan is a software professional with 20 years of experience in the IT industry. She has worked extensively in mobile application development and has led various enterprise mobile enablement initiatives for large organizations and consumer applications for customers around the globe. She is also interested in emerging areas, such as machine learning, IoT, and robotic process automation. She has authored a book with Packt titled *Mobile Machine Learning*. Revathi resides in Chennai and enjoys spending her weekends with her husband and her two lovely daughters.

Jose Miguel is a software engineer who specializes in mobile development, with 5 years of experience. He has an Android associate developer certification from Google. He is also involved in the start-up tech community as a mentor in the OpenLab Peru community, bringing guidance to new Android developers and entrepreneurs who want to gain certain expertise in the mobile world. He resides in Lima, Peru, and enjoys watching movies, reading comics, boxing, traveling, and learning about new cultures and people around the world.

Table of Contents

3

Understanding Data Presentation on Android

4

Managing Dependencies in Android Applications

Part 2 – Domain and Data Layers

5

Building the Domain of an Android Application

6

Assembling a Repository

7

Building Data Sources

Part 3 – Presentation Layer

8

Implementing an MVVM Architecture

9

Implementing an MVI Architecture

10

Putting It All Together

Index

Other Books You May Enjoy

Preface

As an application's code base increases, it becomes harder for developers to maintain existing features and introduce new ones. In this clean architecture book, you'll learn how to identify when and how this problem emerges and how to structure your code to overcome it.

The book starts by explaining clean architecture principles and Android architecture components and then explores the tools, frameworks, and libraries involved. You'll learn how to structure your application in the Data and Domain layers, the technologies that go in each layer, and the role that each layer plays in keeping your application clean. You'll understand how to arrange the code into these two layers and the components involved in assembling them. Finally, we'll cover the Presentation layer and the patterns that can be applied to have a decoupled and testable code base.

By the end of this book, you'll be able to build an application following clean architecture principles and have the knowledge you need to maintain and test the application easily.

Who this book is for

This book is for Android developers who want to learn about managing the complexity of their applications and is also highly recommended for intermediate or advanced Android developers looking for a go-to guide for clean architecture and the integration of various Android technologies. New developers familiar with the fundamentals of Android app development will find this book useful, too.

What this book covers

Chapter 1, *Getting Started with Clean Architecture*, starts by presenting the evolution of Android apps with regards to how business logic was structured, and the problems caused by these approaches. It will then transition to how certain patterns were applied to tackle these issues, revealing other sets of issues. Finally, the concept of clean architecture will be introduce, as well as how its principles can be used to solve some of the problems presented previously.

Chapter 2, Deep Diving into Data Sources, covers what Android tools and frameworks are available to use with regard to the implementation of the data layer and details and expands on the ones that will be used later in the book, such as Kotlin flows and coroutines, Retrofit, Room, and DataStore.

Chapter 3, Understanding Data Presentation on Android, covers what Android tools and frameworks are available to use with regard to the implementation of the presentation layer and will detail and expand on the ones that will be used later in the book, such as Android ViewModel and Jetpack Compose.

Chapter 4, Managing Dependencies in Android Applications, provides a quick overview of dependency injection and how it works. It briefly explores some of the dependency injection tools available for Android development, ending with the Hilt dependency injection framework, about which it goes into a more detailed explanation because it will be used in many of the exercises in the book.

Chapter 5, Building the Domain of an Android Application, describes how to build a domain layer and what components are part of this layer. You will learn about entities and use cases or interactors and what roles they play when it comes to designing the architecture of your application.

Chapter 6, Assembling a Repository, covers the Data layer and the responsibilities this layer has when it comes to managing an application's data, and how it can use the Repository pattern to achieve this.

Chapter 7, Building Data Sources, continues the exploration into the Data layer and some examples of data sources that can be defined in Android. You will learn about using remote data sources to load data from various servers as well as local data sources such as Room and DataStore.

Chapter 8, Implementing an MVVM Architecture, presents the MVVM architecture pattern and how it can be used in an application's presentation layer. You will learn how to use the Android ViewModel and LiveData to build an app with MVVM and integrate use cases into the ViewModel.

Chapter 9, Implementing an MVI Architecture, presents the MVI architecture pattern and how it can be used in an application's presentation layer. You will learn how to use Kotlin flows and Android ViewModel to implement the MVI pattern.

Chapter 10, Putting It All Together, covers the benefits of clean architecture by analyzing an example of an application that implements the concepts and then adding instrumentation tests with Espresso and Jetpack Compose. The introduction of UI tests serves as a good example of how we can inject and change certain behaviors in the application for testing purposes without needing to modify the application's code.

To get the most out of this book

You'll need the Android Studio IDE installed on your computer (version Arctic Fox 2020.3.1 Patch 3 or above) and Java 8 to be installed. Using later versions of Java such as Java 11 might cause errors when building some of the exercises. Knowing how to trigger builds on an emulator or device and Gradle Synchronizations from Android Studio is recommended before attempting the exercises presented in the book.

Software/hardware covered in the book	Operating system requirements
Android SDK 21-32	Windows, macOS, or Linux
Java 8	
Kotlin 1.5.31	

You can expand on the final exercise of the book by optimizing the way the data is loaded, introducing in-memory caches, or integrating new network calls to fetch additional data for the users. You can also improve the instrumentation testing by adding interaction with the list of data and opening new screens and asserting that the correct data is displayed.

If you are using the digital version of this book, we advise you to type the code yourself or access the code from the book's GitHub repository (a link is available in the next section). Doing so will help you avoid any potential errors related to the copying and pasting of code.

Download the example code files

The code bundle for the book is also hosted on GitHub at `https://github.com/PacktPublishing/Clean-Android-Architecture`. If there's an update to the code, it will be updated on the existing GitHub repository.

We also have other code bundles from our rich catalog of books and videos available at `https://github.com/PacktPublishing/`. Check them out!

Code in Action

The Code in Action videos for this book can be viewed at `https://bit.ly/3LqAa30`

Download the color images

We also provide a PDF file that has color images of the screenshots and diagrams used in this book. You can download it here: `https://static.packt-cdn.com/downloads/9781803234588_ColorImages.pdf`

Conventions used

There are a number of text conventions used throughout this book.

`Code in text`: Indicates code words in text, database table names, folder names, filenames, file extensions, pathnames, dummy URLs, user input, and Twitter handles. Here is an example: "Inside the `resources` folder, create a subfolder called `mockito-extensions`. Inside this folder, create a file named `org.mockito.plugins.MockMaker`, and inside this file, add the text `mock-maker-inline`."

A block of code is set as follows:

```
data class User(
    val id: String,
    val firstName: String,
    val lastName: String,
    val email: String
) {

    fun getFullName() = "$firstName $lastName"
}
```

When we wish to draw your attention to a particular part of a code block, the relevant lines or items are set in bold:

```
...
@Composable
fun Screen(viewModel: MainViewModel = viewModel(factory =
MainViewModelFactory())) {
    viewModel.uiStateLiveData.observeAsState().value?.let {
        UserList(uiState = it)
    }
}
...
```

Bold: Indicates a new term, an important word, or words that you see onscreen. For instance, words in menus or dialog boxes appear in **bold**. Here is an example: "Create a new project in Android Studio using an **Empty Compose Activity**."

> Tips or Important Notes
> Appear like this.

Get in touch

Feedback from our readers is always welcome.

General feedback: If you have questions about any aspect of this book, email us at customercare@packtpub.com and mention the book title in the subject of your message.

Errata: Although we have taken every care to ensure the accuracy of our content, mistakes do happen. If you have found a mistake in this book, we would be grateful if you would report this to us. Please visit www.packtpub.com/support/errata and fill in the form.

Piracy: If you come across any illegal copies of our works in any form on the internet, we would be grateful if you would provide us with the location address or website name. Please contact us at copyright@packt.com with a link to the material.

If you are interested in becoming an author: If there is a topic that you have expertise in and you are interested in either writing or contributing to a book, please visit authors.packtpub.com.

Share Your Thoughts

Once you've read Clean Android Architecture, we'd love to hear your thoughts! Scan the QR code below to go straight to the Amazon review page for this book and share your feedback.

https://packt.link/r/180323458X

Your review is important to us and the tech community and will help us make sure we're delivering excellent quality content.

Part 1 – Introduction

In this part, you will become familiar with the notion of clean architecture and the principles it provides. This part also explores the tools, frameworks, and libraries used later in the book.

This part includes the following chapters:

1

Getting Started with Clean Architecture

In this chapter, we'll take you back and show you how a feature would have been implemented in the past while analyzing the potential issues and problems with that approach. Then, we'll look at some key design principles for software development and apply those principles to our legacy examples. After that, we'll cover the evolution of the Android platform and the various libraries and frameworks that have emerged. We'll also see how they can be integrated while adhering to various software design principles.

After that, we'll introduce clean architecture so that we know what our system needs to be improved and what questions we must ask, as developers, so that we can create a robust, scalable, maintainable, and testable application.

In this chapter, we're going to cover the following main topics:

- The architecture of a legacy app
- Software design principles
- Exploring the evolution of Android
- Enter clean architecture

By the end of this chapter, you will know about the evolution of Android development, its architecture, and its design concepts, as well as the concept of clean architecture and how it can be used to build flexible, maintainable, and testable applications.

Technical requirements

For this chapter, you will need Android Studio Arctic Fox 2020.3.1 Patch 3.

The following are the hardware requirements for this chapter:

- Windows:
  - 64-bit Microsoft® Windows® 8/10
  - x86_64 CPU architecture; 2nd generation Intel Core or newer, or an AMD CPU with support for a Windows Hypervisor
  - 8 GB of RAM or more
  - 8 GB of available disk space minimum (IDE + Android SDK + Android Emulator)
  - 1,280 x 800 minimum screen resolution

- Mac:
  - macOS® 10.14 (Mojave) or higher
  - ARM-based chips, or 2nd generation Intel Core or newer with support for Hypervisor.Framework
  - 8 GB of RAM or more
  - 8 GB of available disk space minimum (IDE + Android SDK + Android Emulator)
  - 1,280 x 800 minimum screen resolution

- Linux:
  - Any 64-bit Linux distribution that supports Gnome, KDE, or Unity DE; GNU C Library (glibc) 2.31 or later
  - x86_64 CPU architecture; 2nd generation Intel Core or newer, or AMD processor with support for AMD Virtualization (AMD-V) and SSSE3
  - 8 GB of RAM or more
  - 8 GB of available disk space minimum (IDE + Android SDK + Android Emulator)
  - 1,280 x 800 minimum screen resolution

The architecture of a legacy app

In this section, we will look at how Android applications used to be built in the past and what difficulties developers had with the approach taken.

Before we start analyzing an older application, we must distinguish the architecture and design of an application. To borrow from the construction industry, we can define architecture as a plan for the structure of a building; a design would be a plan to create each part of the building. Translating this into the world of software engineering, we can say that the architecture of an application or a system would be defining a plan that would incorporate the business and technical requirements, while software design deals with integrating all the components, modules, and frameworks into this plan. In an ideal world, you would want to recognize the architecture of an application in the same way you would recognize the architecture of your house.

Now, let's look at the four main components of an Android application:

- **Activities**: These represent the entry points for interacting with the user.
- **Services**: These represent the entry points for having an app run in the background for all kinds of reasons, such as large downloads or audio playback.
- **Broadcast Receivers**: These allow the system to interact with an application for a variety of reasons.
- **Content Providers**: These represent a way for an application to manage application data.

Using and relying on these components created a challenge for developers because an app's architecture became dependent on the Android framework, mainly when it came to implementing unit tests. To understand why this is a problem, let's look at an example of what some older application code would look like. Let's suppose you have been asked to fetch some data from a backend service. The data would be served in the form of JSON through an HTTP connection.

It wasn't uncommon to see a class such as BaseRequest.java, which would execute the request and depend on abstraction in the form of JsonMapper.java, to convert the data from a String into a **Plain Old Java Object** (**POJO**). The following code represents an example of how fetching the data might be implemented:

```
public class BaseRequest<O> {

    private final JsonMapper<O> mapper;

    protected BaseRequest(JsonMapper<O> mapper) {
        this.mapper = mapper;
    }
```

```
public O execute() {
    try {
        URL url = new URL("schema://host.com/path");
        HttpURLConnection urlConnection =
            (HttpURLConnection) url.openConnection();
        int code = urlConnection.getResponseCode();
        StringBuilder sb = new StringBuilder();
        BufferedReader rd = new BufferedReader(new
        InputStreamReader(urlConnection.
            getInputStream()));
        String line;
        while ((line = rd.readLine()) != null) {
            sb.append(line);
        }
        return mapper.convert(new JSONObject
            (sb.toString()));
    } catch (Exception e) {
        ...
    } finally {
        if (urlConnection != null) {
            urlConnection.disconnect();
        }
    }
    return null;
}
}
```

In the `execute` method, we would use `HttpURLConnection` to connect to the backend service and retrieve the data. Then, we would read it into a `String`, which would then be converted into a `JSONObject` and then passed to `JsonMapper` to be converted into a POJO.

The `JsonMapper.java` interface would look something like this:

```
interface JsonMapper<T> {
    T convert(JSONObject jsonObject) throws JSONException;
}
```

This interface represents the abstraction of converting a JSONObject into any POJO.

The use of generics allows us to apply this logic to any POJO. In our case, the POJO should look something like ConcreteData.java:

```
public class ConcreteData {

    private final String field1;
    private final String field2;

    public ConcreteData(String field1, String field2) {
        this.field1 = field1;
        this.field2 = field2;
    }
    public String getField1() {
        return field1;
    }
    public String getField2() {
        return field2;
    }
}
```

The ConcreteData class will be responsible for holding the data we will receive from the backend service. In this case, we just have two String instance variables.

Now, we need to create a concrete JsonMapper.java that will be responsible for converting a JSONObject into ConcreteData:

```
public class ConcreteMapper implements JsonMapper<ConcreteData>
{

    @Override
    public ConcreteData convert(JSONObject jsonObject) {
        return new ConcreteData(jsonObject.optString
            ("field1"), jsonObject.optString("field2"));
    }
}
```

The convert method creates a new ConcreteData object, extracts the data from the JSONObject object, and populates the field1 and field2 values.

Next, we must create a ConcreteRequest.java that will extend BaseRequest and use ConcreteMapper:

```
public class ConcreteRequest extends BaseRequest<ConcreteData>
{

    public ConcreteRequest() {
        super(new ConcreteMapper());
    }

}
```

This class will inherit the execute method from BaseRequest and supply a new ConcreteMapper object so that we can convert the backend data into ConcreteData.

Finally, we can use this in our Activity to execute the request and update our **user interface (UI)** with the result. Here, we have a limitation: we cannot execute long-running operations on the main (UI) thread and we cannot update our views from any other thread except the UI thread. This means that we would need something to help with this. Luckily, Android provides the AsyncTask class, which offers a set of methods for doing work on a separate thread and then processing the results on the main thread. However, we risk creating a context leak (if, for any reason, the Activity object is destroyed, then the garbage collector will not be able to collect the Activity object while AsyncTask is running since Activity has a dependency on AsyncTask) by using an inner AsyncTask class. To circumvent this, the recommended approach is to create a WeakReference for our Activity. This way, if the Activity object is destroyed either by the user or the system, its reference can be collected by the garbage collector.

Now, let's look at the code for our MainActivity:

```
public class MainActivity extends Activity {

    private TextView textView;

    @Override
    protected void onCreate(Bundle savedInstanceState) {
        super.onCreate(savedInstanceState);
        setContentView(R.layout.activity_main);
        this.textView = findViewById(R.id.text_view);
```

```
            new LoadConcreteDataTask(this).execute();
    }

    private void update(ConcreteData concreteData) {
        textView.setText(concreteData.getField1());
    }
}
```

This class is responsible for loading the UI and starting `LoadConcreteDataTask`. The `update` method will then be called by `LoadConcreteDataTask` to show the data in the user interface.

`LoadConcreteDataTask` must be an inner class of `MainActivity`:

```
public class MainActivity extends Activity {
    ...

    private static class LoadConcreteDataTask extends
        AsyncTask<Void, Void, ConcreteData> {
        private final WeakReference<MainActivity>
            mainActivityWeakReference;
        private LoadConcreteDataTask(MainActivity
            mainActivity) {
            this.mainActivityWeakReference = new
                WeakReference<>(mainActivity);
        }

        @Override
        protected ConcreteData doInBackground(Void...
            voids) {
            return new ConcreteRequest().execute();
        }

        @Override
        protected void onPostExecute(ConcreteData
            concreteData) {
            super.onPostExecute(concreteData);
            MainActivity mainActivity =
```

```
                    mainActivityWeakReference.get();
            if (mainActivity != null) {
                mainActivity.update(concreteData);
            }
        }
    }
}
```

In `LoadConcreteDataTask`, we take advantage of the `doInBackground` method, which is executed on a separate thread to load our data and then update our UI in the `onPostExecute` method. We also hold a `WeakReference` to `MainActivity` so that it can be safely garbage collected when destroyed. This also means that we will need to check if the reference still exists before updating the user interface.

The class diagram for the preceding code looks as follows:

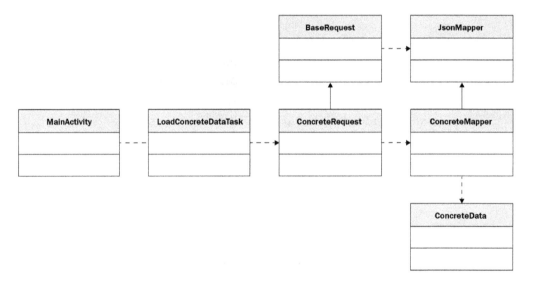

Figure 1.1 – A class diagram for an older Android app

Here, we can see how the dependencies move from `MainActivity` toward the `ConcreteRequest` class, with one exception between `MainActivity` and `LoadConcreteDataTask`, where both classes depend on each other. This is a problem because the classes are then coupled together and making a change to one implies making a change to the other. Later in this chapter, we will look at some principles that can help us avoid such dependencies.

Now that we have an idea of what a legacy application looks like, let's see what issues we may encounter if we follow this path.

Legacy analysis

In this section, we will analyze some of the problems that legacy applications have.

Let's ask ourselves the following questions:

1. What can we unit test?

2. What happens if, instead of showing the value of `field1` from `ConcreteData`, we need to show `field1+field2`?

3. What happens when the requirements for this particular screen change and data needs to be retrieved from another endpoint on top of this one?

4. What happens if we need to introduce caching or SQLite persistence?

5. What happens if another activity needs this particular use case?

Let's answer these questions:

*   **Answer 1**: The answers to all of these questions will come with headaches. The first question is a mix of technical limitations and doubtful design techniques. The technical limitation comes from the fact that the code will execute on the device or an emulator, but we want our unit tests to be executed on our development machines. This is the reason we have the split between the `androidTest` and `test` directories. Theoretically, we can write our unit tests so that they can run on the emulator, but that takes more time and instability. We can now execute these types of tests in the cloud using technologies such as Firebase Test Lab, but that would inevitably cost us money and it's in our interest to avoid taking in such costs. Realistically, we are left with one option and that is to test as much as possible using local unit tests instead of instrumented ones. To solve this, we will need to separate the Android components we use from the Java components.

*   **Answer 2**: The second question produces a similar problem. The easiest choice here would be to put that concatenation into `MainActivity` or add a method into the `ConcreteData` class that will return the concatenated result. But either of these will come with downsides. If we move the concatenation into `MainActivity`, we will put logic that can be unit tested into a class that is very hard and shouldn't be unit tested. If we create a method to concatenate in `ConcreteData`, we risk giving responsibility to this class that it shouldn't have since it's related more to the UI than the actual representation of the JSON itself. What if, in the future, the networking aspect is developed by another team? You would need to rely on that particular team to create this update.

- **Answer 3**: The answer to the third question looks straightforward as well. We must create new concrete implementations for the new data to be added and the associated request. Then, we will either create a separate class that will extend `AsyncTask` or execute both requests in the same `LoadConcreteData` class and then update the UI. If we create a separate `AsyncTask`, then we will need to make the activity responsible for managing the results and balance the two `AsyncTasks`, which again creates a problem concerning testing. If we execute the requests in the same `AsyncTask`, then the responsibility of `AsyncTask` increases, which we may want to avoid.

- **Answer 4**: The fourth question presents us with a new challenge. Let's say we add a new database class that contains all the methods to perform **create, read, update, and delete (CRUD)** operations. Which one of our classes would have a dependency on this class? The choices here would be between the two request classes and `LoadConcreteDataTask`. Here, we run into the same issues that we did in the previous questions. If we used the request classes, we would end up being more responsible for dealing with HTTP connections than handling calls to the database. If we use `LoadConcreteDataTask`, we make the answer to the fifth question even harder.

- **Answer 5**: Based on the previous answers, we notice that a lot of work may end up being moved to the `LoadConcreteDataTask` class. Now, let's imagine that another activity with a completely different UI and a different interpretation of that data will rely on the same use case. One solution is to duplicate `LoadConcreteDataTask` into the new activity. This is not a great idea because a change in the requirements will make the developers change all the tasks. A better approach would be to create a new abstraction that will remove the dependency between `LoadConcreteDataTask` and `Activity`. This would allow us to reuse the same class for both activities. Let's say that the activities would need different types of data for each interpretation. Here, we could follow the `JsonMapper` example and create an interface that would convert `ConcreteData` into a generic type, provide two implementations for each activity, and create the necessary POJOs to convert into.

Another question that can be asked here is, "What amount of work would be necessary to export the business logic into another project?" This is an important question because it highlights how we should structure our code so that it can be reused by others without making it a pain for them to integrate. If we were to answer this, we must first ask, "Where's the business logic?" The answer would probably be `LoadConcreteDataTask`. Can we export that and publish it somewhere where other developers can get it?

The answer is no, because of its dependency on `MainActivity`. This question highlights an important aspect of defining an architecture, namely drawing the boundaries around your components. A component can be defined as the smallest piece of deliverable code. In our case, it would be the equivalent of a module. Now, let's say we were in a place where we could ship out our `LoadConcreteDataTask`. A follow-up question would be, "Would the data be hosted on the same service?" followed by, "Is it in the same JSON format?" Here, we would need to draw a boundary between `LoadConcreteDataTask` and `BaseRequest` and remove such dependencies on how the data is retrieved.

The reason these questions were raised and answered is that all those scenarios have happened in the past and they will all likely happen in the life cycle of an application. We, as developers, tend to answer those questions in our code differently based either on time constraints, the rigor imposed on the team we work in, our ambition to deliver something fast by constantly challenging ourselves, and our experience or the team's experience. The fact that we had the option to make a less desirable solution or to be stuck in a situation where we had to pick between the frying pan or the fire represents a problem. Sometimes, it is good to take a step back from our daily routine, ask ourselves some of these questions, do mind experiments to see how our code may end up in those scenarios, and assess what would happen if that would happen now or 1 or 2 years from now.

A common scenario a lot of Android developers found themselves in was having a lack of businesses investing in testing because it would take too much time and there was a need to go to market. In many of these cases, the apps became harder to maintain over time, so more developers needed to be hired to keep the same productivity as a team compared to when they had fewer developers. When code is written with the notion that it needs to be unit tested, then the way we write that code becomes more rigorous and more maintainable. We start keeping track of how we create instances and separate the things we can test from the things we can't, we apply creational design patterns, and we also shorten the sizes of the methods in our classes, among other things.

We now have an idea of how applications used to be written in the past and the problems that were caused by the approaches that were taken, such as issues with the testability and maintainability of an application due to dependencies on the Android framework. Next, we will look at some design principles that will prove useful in how we write an application.

Software design principles

In this section, we will analyze a set of design principles that are adopted by developers worldwide to improve their systems and can also be applied to Android development. We will mainly focus on the principles defined by *Robert C Martin* (also known as Uncle Bob) for classes and components because they are well suited to Android development.

Based on the examples in the previous section, we understand that our code bases should be maintainable, understandable, and flexible. There is a set of software design principles that we can turn to for help when we develop classes or components. Think of a component as the minimum amount of code that can be released as part of a system. In Android, you can view them as individual modules. They don't necessarily need to be modules, but they can be organized as if they are.

SOLID principles

These are some of the most known design principles. The name is an acronym for a set of design principles that were collected by *Robert C Martin*. These principles are as follows:

- Single responsibility principle
- Open-closed principle
- Liskov substitution principle
- Interface segregation principle
- Dependency inversion principle

Let's look at these principles in detail:

- **Single Responsibility Principle**: This states that a class should have one responsibility or one reason to change. Looking at our example, let's suppose someone makes a change to the `BaseRequest` class to change how the HTTP request is executed. Let's assume that we now have two different AsyncTasks that will load the data. Both of these will be impacted by the change in the `BaseRequest` class. A solution would be to delegate the execution of the request to different classes for each particular use case. This would also allow developers to work on different features related to backend communication without changing the same source file.

- **Open-Closed Principle**: This states that a class should be open for extension and closed for modification. Thinking back to our example, this principle would answer the question, "What would happen if an activity requires this particular use case?" The abstractions we discussed in how to answer that question would serve as a good example of implementing this principle.

- **Liskov Substitution Principle**: This states that a parent class should be replaced by a child class without changing the behavior of the system. An example of this principle is if you have a class called `Bird` and a sub-class called `Duck`. If you are using references of `Bird` in your code and substitute those usages with `Duck`, then your code should remain unchanged. A famous example of a violation of this principle is having a `Rectangle` class with two members named `width` and `height` and a sub-class named `Square`. In reality, a square is a rectangle, but our modeling of a square wouldn't be a rectangle because the rules in `Square` would mean that the width and height will always have to be the same. If you were to swap these two dependencies, then your code would break.

- **Interface Segregation Principle**: This states that we should avoid using large interfaces and instead break them up into smaller interfaces. The idea here is that code shouldn't depend on methods it doesn't use. An example of this is defining interfaces whose methods don't need to be implemented. A good example of this is the approach that's taken in Android user interfaces by separating `OnClickListener`, `OnLongClickListener`, and `OnTouchListener`.

- **Dependency Inversion Principle**: This states that we should depend on abstractions rather than concretions. The idea here is to depend as much as possible on abstract classes and interfaces. This can be very difficult to achieve considering that we rely on concretions a lot of the time. Here, we should identify parts of the code that are constantly developed and subject to change and introduce layers of abstractions between our code and these classes. A good way to protect against this is through dependency injection frameworks such as Dagger and Hilt, which generate factories to create volatile components.

SOLID principles are used across the **object-oriented programming** (**OOP**) field to create applications that are flexible and able to incorporate new features and requirements. The principles that follow represent an expansion of SOLID.

Component cohesion principles

We can define cohesion by how well the classes in a component belong together or what classes belong in a certain component. In the past, components were assembled based on the context without any particular guiding principle. This would cause issues such as a change in the dependencies of a component triggering a change in the dependants of this component, without this having any relevance to the dependants.

The three principles are as follows:

- **Reuse/Release Equivalence Principle** (REP): This states that we group classes in a component that can be released together. In Android development, this would translate to making sure that every module you create should be able to be published and used by other developers.

- **Common Closure Principle** (CCP): This states that components should have one reason to change. This principle is an application of the single responsibility principle for components.

- **Common Reuse Principle** (CRP): This states that a component should only have classes that should be used together. This represents the interface segregation principle for your component. In Android, this would mean that you should make sure that the users of your Android modules depend on all your classes in the module, not just some.

When these principles are incorporated, they end up conflicting with each other. REP and CCP tend to make components bigger, while CRP tends to make them smaller. The idea is to always match the current requirements of the application and find the middle ground between these principles. After that, you should constantly monitor how new requirements would affect this middle ground.

Now that we've seen how SOLID can be applied to building a particular component through the component cohesion principles, let's learn how to manage a set of components.

Component coupling principles

These principles deal with how to manage the relationships between our components in an Android application. In Android, this would be represented by how to manage the Gradle dependencies between different modules. The principles are as follows:

- **Acyclic Dependencies Principle**: This states that we should avoid cyclic dependencies between components. Applying this to Android would mean that the dependencies that our modules have most not be cyclical (for example, module A depends on module B, which depends on module A). Fortunately, this rule is currently enforced by the build system, which doesn't allow cyclical dependencies. A solution to this would be to create a new module in which we apply the dependency inversion principle and make one of the modules depend on the abstraction and create the implementation in the second module. If this is not possible, we can create a new module that can depend on both existing modules. An example of this can be seen in the following diagram:

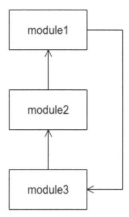

Figure 1.2 – Cyclic module dependency

- **Stable Dependencies Principle**: This states that less stable modules should depend on more stable modules. A component's stability is defined as the ratio between outgoing dependencies (dependency on other components) and the total number of dependencies. The closer the number is to 0, the more stable a component becomes. This means that stable components should avoid having changes made because this will cause potential issues for the components that depend on the stable ones. One solution to avoid the dependencies between stable components and volatile components would be using abstract components. These are components that will contain nothing but abstractions.

- **Stable Abstractions Principle**: This states that components that are likely to change should be more concrete and that stable components should be more abstract. This principle represents an application of the open-closed principle. We would want our high-level architecture decisions to be flexible enough to be changed without having to modify existing source code. We can achieve this using abstract classes. The abstractness of a component is defined as the ratio between the number of abstract classes and interfaces inside a component and the total number of classes in the component. The closer to 1 the value gets, the more abstract the component becomes. A component with 0 stability and 0 abstractness represents a **zone of pain** because it is very hard to change. A component with 1 stability and 1 abstractness is called a **zone of uselessness** because we have an independent component with no implementations. The aim is to get as many components as possible in either the 0 stability and 1 abstractness or 1 stability and 0 abstractness range.

With that, we have looked at some of the key design principles that should help us tackle problems that we face while developing an application. The SOLID principles show us how we should structure our code into classes, while the component cohesion principles and component coupling principles show us how we should structure our classes into separate modules, as well as how we should establish the relationships between those modules. In the next section, we will see how these principles lead to the evolution of the Android platform and what an application may look like now.

Exploring the evolution of Android

In this section, we will look at key releases and changes that have been made to the Android framework and supporting libraries that have shaped the development of applications and how applications have evolved because of these changes.

We started by looking at an example of what the code in an older Android application looked like before looking at the design principles we should incorporate into our work. Now, let's see how the Android framework evolved and how some of our questions from the beginning have been answered. We will analyze some of the newer libraries, frameworks, and technologies that we can incorporate into an Android application.

Fragments

The introduction of fragments was meant to solve important issues developers were facing – that is, the activity code would become too big and hard to manage. They were released on Android Honeycomb, which was an Android release that only targeted tablets. The introduction of fragments was also meant to solve the issue of having different displays for activities in landscape versus activities in portrait. Fragments are meant to control portions of an activity's user interface.

Another improvement fragments brought was the ability to change and replace fragments at runtime. There was even a separate back stack for Fragments that the activity would be responsible for. This comes at a couple of costs: the life cycle of the fragment was even more complex than the life cycle of the activity, where you would have fragments that had their views destroyed but the fragments themselves weren't. Another cost was the communication between two fragments. If you needed to update the user interface being handled by Fragment1 because of a change in Fragment2, you would need to communicate through the activity. This meant that every time a Fragment needed to be reused by a different activity, then the activity would be forced to adapt to this:

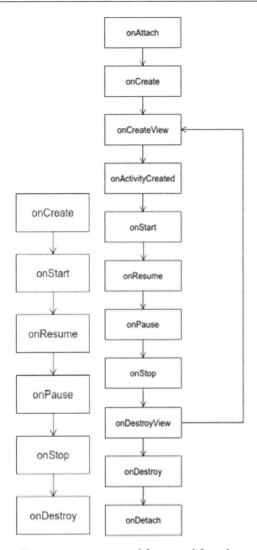

Figure 1.3 – Activity and fragment life cycle

In the preceding figure, we can see the difference between the lifecycle of activities and the lifecycle of fragments. We can observe how fragments have their own internal lifecycle for managing the views that they display between the onCreateView method and onDestroyView methods. This is often the reason why in many applications, you will see these methods used to load data and on the opposite site unsubscribing from any operations that might trigger a change in the user interface.

The Gradle build system

Initially, Android development used the Eclipse IDE and Ant as its build system. This came with certain limitations for applications. Things such as flavors were not available at the time. The release of Android Studio, along with the Gradle build system, provided new opportunities and features. This allows us to write extra scripts and easily integrate plugins and tools, such as performance monitoring of an application, Google Play services, Firebase Crashlytics, and more. This is often done through ".gradle" files. These files are written in a language called Groovy. Another improvement that was added was the usage of the ".gradle.kts" extensions, where we can provide the same configurations using the Kotlin language. The following code shows what the build.gradle file for a module looks like:

```
plugins {
    id 'com.android.application'
}
android {
    compileSdk 31
    defaultConfig {
        minSdk 21
        targetSdk 31
        versionCode 1
        versionName "1.0"
    }
    buildTypes {
        release {
        }
    }
    compileOptions {
    }
}
dependencies {
    implementation ""
}
```

In the `plugins` section, we can define external plugins that will provide certain methods and scripts that our project can use. Examples include annotation processing plugins, the `Parcelize` plugin, and Room plugins. In this case, the `com.android.application` plugin offers us the `android` configuration, which we can then use to specify the app version, what Android versions we want the app to be accessible from, various compilation options, and configurations for how the app should be built for the end user. In the `dependencies` section, we specify which external libraries we want to add to the project.

Networking

Quite a few popular networking libraries have emerged, mainly in the open sourcing community. A large proportion of the applications in Google Play rely on HTTP communication and a large proportion of them use JSON data. With the addition of networking libraries, JSON serialization/deserialization to POJOs also became adopted. What this means for developers is that the communication with the backend is simplified – we no longer need to concern ourselves with how the actual communication is done; we only point to where we want the data from and provide the models that are required for this communication. The libraries will take care of the rest. Some of the most popular libraries include Volley and Retrofit. In terms of object serialization, we have libraries such as Moshi and GSON.

Humble objects

Because activities and fragments are difficult to unit test, the code inside them needed to be split into testable sections and untestable sections. Because of this necessity, two patterns emerged: **Model View Presenter** (**MVP**) and **Model View ViewModel** (**MVVM**). Sometimes, these patterns are referred to as architecture patterns. This shouldn't be confused with the entire architecture of the app. The idea is to turn activities and fragments into humble objects with no logic, keep the references to the user interface objects, and shift the logic into the presenter and ViewModel, which we can write unit tests for. We will focus more on the particularities of each in *Chapter 8, Implementing an MVVM Architecture*.

Functional paradigms

Just like objected-oriented languages have adopted paradigms from functional programming, so has the Android development world in the form of RxJava. Functional programming works on the premise that programs are built from composing functions rather than imperative statements such as the ones in Java. RxJava is a library that allows developers to implement event-driven applications. It offers observables (for emitting data) and subscribers (for subscribing to that data). What made this library appealing to developers was how it deals with threading. Let's assume you wanted an operation to be executed on a separate thread, and then you wanted to transform your data – all you need to do here is invoke the data you want, apply mapping functions, and then subscribe to get the final result. The added benefit is that you can chain different operations, have them processed, and get the result with all of the operations. All of this removes the need for creating and managing different AsyncTasks or threads.

Kotlin adoption

RxJava introduced some aspects of functional programming. Its adoption and transition into Kotlin programming has added others. One of the most important is the concept of mutability. In Java, all variables are mutable unless they're declared otherwise through the `final` keyword. In Kotlin, all the variables must have their mutability declared. Why is this important? Because of multi-threading. If you had an application where multiple threads were executed at the same time and they all interacted with the same object, you would end up in a situation where you would either modify the same value at the same time or create deadlocks in which a thread would wait for another thread to release a resource, but the second thread would need access to a resource that the first thread is currently holding. This introduction helps developers aim for a greater degree of immutability, which would increase thread safety because immutable variables are thread-safe. Lambdas represent another great feature of Kotlin that allows boilerplate code to be reduced when you're dealing with callbacks. Other benefits of the adoption of Kotlin include that you can remove boilerplate code by introducing data classes, which represent POJOs, and introducing sealed classes, which allow developers to define enum-like structures that can carry data.

Dependency injection

Dependency injection represents the decoupling of object invocation and object creation. Why is this important? Mainly because of testing. It's easier to write unit tests for classes that have their dependencies injected rather than adding extra responsibilities, such as creating new instances for all of the dependencies in that class. Another benefit is in situations where we depend on abstractions. If we have a dependency on an abstraction, we can easily switch between different implementations, depending on different circumstances. Several libraries have emerged to tackle this issue: Dagger, Koin, and Hilt. Dagger is more of a general library that is not only Android applicable, but also applicable for other Java-based platforms. It aims to manage our dependencies using components and modules. Components are responsible for how the dependencies are managed, while modules are responsible for providing the appropriate dependencies. It relies on annotation processors, which generate the necessary code that will be responsible for managing our dependencies. Koin is what's referred to as a service locator library. It keeps a collection of all the dependencies and when a particular dependency is required, it will look it up and provide it. Koin is an Android-specific library, and it provides support for injecting specific Android dependencies. Hilt is the newest of these libraries and it is built on top of Dagger. It removes the boilerplate code that was required for Dagger and provides support for Android dependencies as well.

Android architecture components

This is represented by a set of libraries that help developers make their apps scalable, testable, and maintainable. These libraries affect components that deal with activity and fragment life cycles, persisting data, background work, and UIs. Here, we have seen the introduction of concepts such as life cycle owners (such as activities and fragments), the Android ViewModel, and LiveData. These are meant to solve problems developers had with managing the state of a life cycle owner when it's destroyed and recreated by the system. It puts the logic that, in the past, was handled by the life cycle owners and delegated to the Android ViewModel. The combination of the Android ViewModel and LiveData has helped developers implement the MVVM pattern, which is also life cycle aware. This means that developers no longer have to concern themselves with stopping a background task when the life cycle owner is destroyed.

The introduction of Room means that developers no longer have to deal with interacting with the SQLite framework, which caused a lot of boilerplate code to be written to define tables and various queries. Developers no longer need to deal with the SQLite interaction and the many dependencies that come with it; instead, they can focus on creating their own models and providing the abstractions for what needs to be queried, deleted, updated, and deleted; Room will take care of the actual implementations. DataStore does for SharedPreferences what Room does for SQLite. This is for when we want to store data in key-value pairs instead of using an entire table. DataStore provides two options for storing data: safely typed data and no type safety data.

With the addition of these new persistence libraries, the Repository pattern was adopted. The idea behind this pattern is to create a class that will interact with all the data sources we have in our application. As an example, let's imagine we have some data we will need to fetch from our backend that will then need to be stored locally in case we want the user to view it offline. Our repository would be responsible for fetching the data from the network class and then storing it using the persistence class. The repository would sit in between the local and remote classes and the classes that would want access to that data.

Regarding the UI, we now have access to view binding and data binding. Both of these deal with how activities and fragments deal with the views that are declared in our XML layout files. View binding generates references for each view we defined in our XML. This solves an issue that developers would have in the past where a view would be deleted from your XML file, but your application would still run because of another view with the same name in another file. This would cause crashes in the past because the findViewById function would return null. With view binding, we know at compile time what views we have in our hierarchy and what views we don't. Data binding allows us to bind our views to data sources. For example, we can bind a TextView in our XML file directly to a field in our source code. This approach tends to work well with the MVVM pattern, in which the ViewModel updates certain fields that are bound by views in our XML. This would update what the view would display without it interacting with the activity.

Coroutines and flows

Coroutines came as a feature of the Kotlin language. The idea behind coroutines is to execute data asynchronously in a very simplified manner. We no longer have to create threads or AsyncTasks (which have been deprecated) and manage concurrency because it's managed under the hood. Other features include that it's not bound to a particular thread, and it can be suspended and resumed. Flows represent an extension of coroutines where we can have multiple emissions of data, such as RxJava, providing similar benefits.

Jetpack Compose

This allows developers to build UIs directly in Kotlin without the use of XML files through composable functions. This removes the amount of code that needs to be written for building your UI. Compatibility with the other Android architecture component libraries is provided, allowing for easier integration into your application. The following is an example of what Compose looks like:

```
class MainActivity : ComponentActivity() {
    override fun onCreate(savedInstanceState: Bundle?) {
        super.onCreate(savedInstanceState)
        setContent {
            ExampleTheme {
                Surface {
                    ExampleScreen()
                }
            }
        }
    }
}


@Composable
fun ExampleScreen() {
    Column(modifier = Modifier.padding(16.dp)) {
        TextField(
            value = "",
            onValueChange = {
                // Handle text change
            },
            label = { Text("Input") }
        )
        Text(text = "Example text")
        Button(onClick = {
            // Handle button click
        }) {
            Text(text = "Button")
        }
    }
}
```

In this example, we can see a screen that contains an input field, some text that displays Example Text, and a button with the text Button. The layout of the screen is defined as a function annotated with the @Compose annotation. This content is then set in an activity through the setContent method, where a theme is provided. We will expand on how Jetpack Compose works later in this book.

Now, let's look at what our example code from the *The architecture of a legacy app* section will look like after we transition it through some of the aforementioned Android frameworks and updates. All our code will now be migrated to Kotlin. We will be using libraries such as Retrofit and Moshi for networking and JSON serialization and Hilt for dependency injection, as well as ViewModel, LiveData, and Compose for the UI layer. We will discuss how these libraries work in the following chapters.

The ConcreteData class will look this:

```
@JsonClass(generateAdapter = true)
data class ConcreteData(
    @Json(name = "field1") val field1: String,
    @Json(name = "field1") val field2: String
)
```

The ConcreteData class is now a Kotlin data class and will use the Moshi library for JSON conversion. Next, let's see what our HTTP request will look like when we use something such as Retrofit to handle our HTTP communication:

```
interface ConcreteDataService {

    @GET("/path")
    suspend fun getConcreteData(): ConcreteData
}
```

Because we use Retrofit and OkHttp, we only need to define the template for the endpoint we want to connect to and the data we want; the libraries will handle the rest. The suspend keyword will come in handy for Kotlin flows.

Now, let's define a repository class that will be responsible for invoking this HTTP call on a separate thread:

```
class ConcreteDataRepository @Inject constructor(private val
concreteDataService: ConcreteDataService) {

    fun getConcreteData(): Flow<ConcreteData> {
```

```
    return flow {
        val fooList = concreteDataService.
            getConcreteData()
        emit(fooList)
    }.flowOn(Dispatchers.IO)
  }
}
```

ConcreteDataRepository will have a dependency on ConcreteDataService, which it will call to fetch the data. It will be responsible for retrieving the data on a separate thread by using Kotlin flows. The constructor will be annotated with the @Inject annotation because we are using Hilt, which will inject ConcreteDataService into ConcreteDataRepository.

Now, let's create a ViewModel that will depend on the repository to load the appropriate data:

```
@HiltViewModel
class MainViewModel @Inject constructor(private val
concreteDataRepository: ConcreteDataRepository) :
    ViewModel() {

    private val _concreteData = MutableLiveData
        <ConcreteData>()
    val concreteData: LiveData<ConcreteData> get() =
        _concreteData

    fun loadConcreteData() {
        viewModelScope.launch {
            concreteDataRepository.getConcreteData()
                .collect { data ->
                    _concreteData.postValue(data)
                }
        }
    }
}
```

`MainViewModel` will then use `ConcreteDataRepository` to retrieve the data, subscribe to the result, and post the result in `LiveData`, which `MainActivity` will subscribe to.

Now, let's create `MainActivity`:

```
@AndroidEntryPoint
class MainActivity : ComponentActivity() {
    override fun onCreate(savedInstanceState: Bundle?) {
        super.onCreate(savedInstanceState)
        setContent {
            Screen()
        }
    }
}


@Composable
fun Screen(mainViewModel: MainViewModel = viewModel()){
    mainViewModel.loadConcreteData()
    UpdateText()
}


@Composable
fun UpdateText(mainViewModel: MainViewModel = viewModel()) {
    val concreteData by mainViewModel.concreteData.
        observeAsState(ConcreteData("test", "test"))
    MessageView(text = concreteData.field1)

}


@Composable
fun MessageView(text: String) {
    Text(text = text)
}
```

`MainActivity` is now written using Jetpack Compose. It will trigger the data load when the screen is created and then subscribe to `LiveData` from `ViewModel`, which will update the text on the screen when the data is loaded.

Since we are using Hilt for dependency injection, we will need to define our external dependencies in a module, as follows:

```
@Module
@InstallIn(SingletonComponent::class)
class ApplicationModule {

    @Singleton
    @Provides
    fun provideHttpClient(): OkHttpClient {
        return OkHttpClient
            .Builder()
            .readTimeout(15, TimeUnit.SECONDS)
            .connectTimeout(15, TimeUnit.SECONDS)
            .build()
    }
}
```

First, we must provide the `OkHttp` client, which is used to make the HTTP requests.

Next, we will need to provide the JSON serialization:

```
@Module
@InstallIn(SingletonComponent::class)
class ApplicationModule {
    ...
    @Singleton
    @Provides
    fun provideConverterFactory(): MoshiConverterFactory =
MoshiConverterFactory.create()
}
```

We are using the Moshi library for JSON serialization, so we will have to provide a Factory that will be used by Retrofit for JSON conversion.

Next, we need to provide a Retrofit object:

```
@Module
@InstallIn(SingletonComponent::class)
class ApplicationModule {
```

```
...
@Singleton
@Provides
fun provideRetrofit(
    okHttpClient: OkHttpClient,
    gsonConverterFactory: MoshiConverterFactory
): Retrofit {
    return Retrofit.Builder()
        .baseUrl("schema://host.com")
        .client(okHttpClient)
        .addConverterFactory(gsonConverterFactory)
        .build()
}
}
```

The Retrofit object will need a base URL that will act as the host for our backend service, OkHttpClient, and the JSON converter factory, which were provided earlier.

Finally, we will need to provide the template we defined previously:

```
@Module
@InstallIn(SingletonComponent::class)
class ApplicationModule {

    @Singleton
    @Provides
    fun provideConcreteDataService(retrofit: Retrofit):
        ConcreteDataService =
            retrofit.create(ConcreteDataService::class.java)
}
```

Here, we will use Retrofit to create an instance of ConcreteDataService that will be injected into ConcreteDataRepository by Hilt.

Finally, we need to initialize Hilt in the Application class:

```
@HiltAndroidApp
class MyApplication : Application()
```

This code represents a 10-year jump in time when it comes to Android development. Going back to the questions we asked for the initial example in the Legacy analysis section, we can see that we answered quite a few. If we want to introduce persistence into the application, we now have a repository that can manage that for us. We also have a lot of classes that can be individually unit tested because of the introduction of Hilt and because we have delimited separated from the Android framework dependencies. We have also introduced flows, which allow us to manipulate and handle the data in case we need to connect to multiple sources and handle multi-threading more easily. The introduction of Kotlin and Retrofit also allowed us to reduce the amount of code. If we were to make a diagram of this, it would look as follows:

Figure 1.4 – A class diagram for a newer Android application

Here, we can see that the dependencies between the classes go from one direction to the other, which is another positive. The introduction of Retrofit saved us a lot of hassle when dealing with HTTP requests. But an issue remains with regards to how `ConcreteData` is handled. We can see that it travels from `ConcreteDataService` into `MainActivity`. Imagine if we wanted to provide the data from a different URL with a different POJO representation. This means that all of the classes will have to be changed to accommodate for this. This violates the single responsibility principle because the `ConcreteData` class is used to serve multiple actors in our application. In the next section, we will try to seek a solution to this problem and address ways to properly structure our classes and components.

With that, we have explored the evolution of the Android platform and tools, what an application may look like using the latest tools and libraries, and how this evolution solved many problems developers had in the past. However, we still haven't solved all of them. In the next section, we will talk about the concept of clean architecture and how we can use it to make our application flexible and more adaptable to changes.

Enter clean architecture

In this section, we will discuss the concept of clean architecture, the problems it solves, and how it can be applied to an Android application.

Architecture can be viewed as the high-level solution that's required to build a system that can solve business and technical requirements. The goal should be to keep as many options on the table for as long as we can. From an Android development perspective, we've seen the platform grow a lot, and to balance the new changes that have been added to the platform with the addition of new features for our application and its maintenance, we will need to give our application a very good foundation so that it will adapt to changes. A common approach to architecture in Android development was the layered architecture, where apps would be split into three layers – the user interface, domain, and data layers. The problem here was that the domain layer depended on the data layer, so when the data layer changed, the domain layer needed to change too.

Clean architecture represents an integration of multiple types of architecture that provide independence from frameworks, user interfaces, and databases, as well as being testable. The shape resembles that of an onion, where dependencies go toward the inner layers. These layers are as follows:

- **Entity Layer**: This layer is the innermost layer and is represented by objects that hold data or business-critical functions.

- **Use Case Layer**: This layer implements the business logic of the system.

- **Interface Adapter Layer**: This layer is responsible for converting the data between the frameworks and drivers and the use case. This will hold components such as ViewModels and presenters, as well as various converters that are responsible for converting network and persistence-related data into entities.

- **Frameworks and Drivers Layer**: This layer is the outermost layer and is comprised of components such as activities, fragments, networking components, and persistence components.

Let's consider a scenario: you've recently been hired by a start-up company as their first Android engineer. You have been given a basic idea of what the app that you've been asked to develop should do, but there isn't anything too concrete; the user interface has not been finalized, the teams working on the backend are new themselves, and there isn't anything too concrete on their side either. What you do know is a set of use cases that specify what the app does: log into a system, load a list of tasks and add new tasks, delete tasks, and edit existing tasks. The product owner tells you that you should work on something while using mock data so that they can get a feel of the product and consult with the user interface and user experience teams to discuss improvements and modifications.

You are faced with a choice here: you can build the product that's been requested by the product owner as fast as possible and then constantly refactor your code for each new integration and the change in requirements, or you can take a little bit more time and factor in the future changes that will come into your approach. If you were to take the first approach, then you would find yourself in a situation where many developers found themselves, which is to go back and change things properly. Let's assume you chose the second approach. What would you need to do then? You can start decoupling your code into separate layers. You know that the UI will change, so you will need to keep it isolated so that when it is changed, the change will only be isolated to that particular section. Often, the UI is referred to as the presentation layer.

Next, you want to decouple the business logic. This is something specific to processing the data that your app will use. This is often done in the domain layer. Finally, you want to decouple how the data is loaded and stored. This will be the part where you deal with integrating libraries such as Room and Retrofit and it's often called the data layer. Because the requirements aren't definitive yet, you also want to decouple how you want to handle use cases so that if a use case changes, you can protect the others from that change. If you were to rotate the class diagram from *Figure 1.4*, you would see a layered approach to this example.

As we've mentioned previously, the fact that `ConcreteData` shows up in all the classes in our example is not a good idea. This is because, at the end of the day, the fact that we chose Retrofit and Moshi shouldn't impact the rest of the application. This is similar if it was the opposite way around and the activity or `ViewModel` would've done the same. At the end of the day, the way we choose to implement our UI or what networking library we should use represent details. Our domain layer shouldn't be impacted by any of these choices.

What we are doing here is establishing boundaries between the components in our system so that a change in a component doesn't impact a change in another component. In Android, even if we use the latest libraries and frameworks, we should still make sure that our domain is still protected by changes in those frameworks. Going back to the start-up example, and assuming you've chosen to decouple your components and pick the appropriate boundaries, after many demos and iterations, your company decides to hire additional developers to work on new, separate features. If those developers follow the guidelines you've set up, they can work with a minimal level of overlap.

The recommendation from Android development documentation is to take advantage of modules. One of the arguments is that it improves build speed because when you work on a certain module, it won't rebuild the others when you build the application – it caches them instead. Splitting your application into multiple modules serves another purpose.

Let's go back to the start-up. Things are going great and people love your product, so your company decides to open your APIs for other businesses to integrate into their systems. Your company also wants to provide an Android library so that it's easier for businesses to access your APIs. You already have this logic integrated into your application; you just need to export it. What features do you want to export? All? None? Do they want to persist data locally? Do they want some of the UI or not? If your modules were split with proper boundaries, then you would be able to accommodate all of those features. What we want to do is have a system where we can easily plug things in and easily plug them out.

Transitioning our previous example to this approach, we would have something like this. The `ConcreteData` class and `ConcreteDataService` would remain the same:

```
@JsonClass(generateAdapter = true)
data class ConcreteData(
    @Json(name = "field1") val field1: String,
    @Json(name = "field1") val field2: String
)


interface ConcreteDataService {

    @GET("/path")
    suspend fun getConcreteData(): ConcreteData
}
```

Now, we will need to isolate the Retrofit library and create the interface adapter for it. But to do that, we will need to define our entity:

```
data class ConcreteEntity(
    val field1: String,
    val field2: String
)
```

It looks like it's a duplicate of `ConcreteData`, but this is a case of fake duplication. In reality, as things evolve, the two classes may contain different data, so they will need to be separated.

To isolate the Retrofit call, we need to invert the dependency of our repository. So, let's create a new interface that will return `ConcreteEntity`:

```
interface ConcreteDataSource {
```

```
    suspend fun getConcreteEntity(): ConcreteEntity
}
```

In our implementation, we will invoke the Retrofit service interface:

```
class ConcreteDataSourceImpl(private val concreteDataService:
ConcreteDataService) :
    ConcreteDataSource {

    override suspend fun getConcreteEntity():
        ConcreteEntity {
        val concreteData = concreteDataService.
            getConcreteData()
        return ConcreteEntity(concreteData.field1,
            concreteData.field2)
    }
}
```

Here, we have invoked `ConcreteDataService` and then converted the network model into an entity.

Now, our repository will change into the following:

```
class ConcreteDataRepository @Inject constructor(private val
concreteDataSource: ConcreteDataSource) {

    suspend fun getConcreteEntity(): ConcreteEntity {
        return concreteDataSource.getConcreteEntity()
    }
}
```

`ConcreteDataRepository` will depend on `ConcreteDataSource` to avoid the dependencies on the networking layer.

Now, we need to build the use case to retrieve `ConcreteEntity`:

```
class ConcreteDataUseCase @Inject constructor(private val
concreteDataRepository: ConcreteDataRepository) {

    fun getConcreteEntity(): Flow<ConcreteEntity> {
```

```
        return flow {
            val fooList = concreteDataRepository.
                getConcreteEntity()
            emit(fooList)
        }.flowOn(Dispatchers.IO)
    }
}
```

`ConcreteDataUseCase` will depend on `ConcreteDataRepository` to retrieve the data and emit it using Kotlin flows.

Now, `MainViewModel` will need to be changed to invoke the use case. To do so, it will use the `field1` object from `ConcreteEntity`:

```
@HiltViewModel
class MainViewModel @Inject constructor(private val
concreteDataUseCase: ConcreteDataUseCase) :
    ViewModel() {

    private val _textData = MutableLiveData<String>()
    val textData: LiveData<String> get() = _textData

    fun loadConcreteData() {
        viewModelScope.launch {
            concreteDataUseCase.getConcreteEntity()
                .collect { data ->
                    _textData.postValue(data.field1)
                }
        }
    }
}
```

`MainViewModel` will now depend on `ConcreteDataUseCase` and retrieve `ConcreteEntity`, where it will extract `field1`. This will then be set in `LiveData`.

`MainActivity` will be updated to use the `textData` object from `MainViewModel`:

```
@AndroidEntryPoint
class MainActivity : ComponentActivity() {
    override fun onCreate(savedInstanceState: Bundle?) {
```

```
        super.onCreate(savedInstanceState)
        setContent {
            Screen()
        }
    }
}

@Composable
fun Screen(mainViewModel: MainViewModel = viewModel()){
    mainViewModel.loadConcreteData()
    UpdateText()
}

@Composable
fun UpdateText(mainViewModel: MainViewModel = viewModel()) {
    val text by mainViewModel.textData.
        observeAsState("test")
    MessageView(text = text)

}

@Composable
fun MessageView(text: String) {
    Text(text = text)
}
```

With that, `MainActivity` has been updated to use `LiveData`, which emits a `String` instead of a `ConcreteData` object.

Finally, the Hilt module will be updated as follows:

```
@Module
@InstallIn(SingletonComponent::class)
class ApplicationModule {
    ...
    @Singleton
    @Provides
    fun provideHttpClient(): OkHttpClient {
```

```kotlin
        return OkHttpClient
            .Builder()
            .readTimeout(15, TimeUnit.SECONDS)
            .connectTimeout(15, TimeUnit.SECONDS)
            .build()
    }


    @Singleton
    @Provides
    fun provideConverterFactory(): MoshiConverterFactory =
        MoshiConverterFactory.create()


    @Singleton
    @Provides
    fun provideRetrofit(
        okHttpClient: OkHttpClient,
        gsonConverterFactory: MoshiConverterFactory
    ): Retrofit {
        return Retrofit.Builder()
            .baseUrl("schema://host.com")
            .client(okHttpClient)
            .addConverterFactory(gsonConverterFactory)
            .build()
    }


    @Singleton
    @Provides
    fun provideCurrencyService(retrofit: Retrofit):
        ConcreteDataService =
        retrofit.create(ConcreteDataService::class.java)


    @Singleton
    @Provides
    fun provideConcreteDataSource(concreteDataService:
        ConcreteDataService): ConcreteDataSource =
        ConcreteDataSourceImpl(concreteDataService)
}
```

Here, we can see that `ConcreteDataUseCase` just invokes `ConcreteDataRepository`, which just invokes `ConcreteDataSource`. You may be wondering why this boilerplate is necessary. In this case, we have a bit of fake duplication. As the code grows, `ConcreteDataRepository` may connect to other data sources, and `ConcreteDataUseCase` may need to connect to multiple repositories to combine the data. The same can be said about `ConcreteData` and `ConcreteEntity`. Another benefit of this approach is the imposition of more rigor when it comes to development, and it creates consistency.

Let's look at the following diagram and see how it compares to *Figure 1.4*:

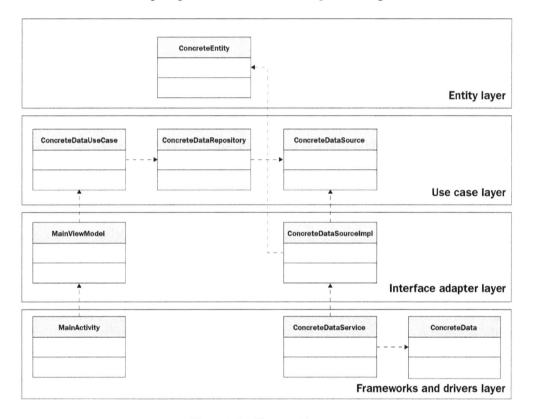

Figure 1.5 – Clean architecture

If we look at the top row, we will see the use case and the entity. We can also see that the dependencies go from the classes at the bottom toward the classes at the top, similar to how the dependencies go from the outer layers toward the inner layers here. A difference you may have noticed is that our example doesn't mention the usage of modules. Later in this book, we will explore how to apply clean architecture to multiple modules and how to manage them.

We are now back in the start-up, and you started working on the application, where you have defined a few entities and use cases and have put a simple UI in place. The product owner has asked you to deliver a demo with some mock data for tomorrow. What can you do? You can create a new implementation of your data source and plug in some mock objects that you can use to satisfy the conditions for the demo. You show the demo of the application and you receive some feedback about your UI. This means you can change your activities and fragments to render the data appropriately, and this won't impact any of the other components. What would happen if the use case were to change? In that situation, this would propagate into the rest of the other layers. This depends on the change, though, but this scenario is to be expected in this situation.

Summary

In this chapter, we explored what an Android app used to look like and all the problems developers would face at the time. We've looked at some of the most important software design principles, such as SOLID, to get a better understanding of how to improve our code and how these principles helped the Android platform evolve. We also looked at the adoption of a new programming language that came with the introduction of new software paradigms, the addition of event-based libraries and frameworks, the introduction of architecture components to help developers write more testable applications, and a new way to build user interfaces. Finally, we introduced clean architecture, which helps us build maintainable, testable, and more independent applications. We looked at all of these changes through a small example, where we saw them transition from what they may have looked like in 2010 to what they may look like now.

In the next chapter, we will deep dive into the libraries that are required for loading, storing, and managing data on Android. We will combine them to build an app using clean architecture.

2
Deep Diving into Data Sources

In this chapter, we will study some of the popular libraries and frameworks used for retrieving and managing data on Android and how to do this without blocking the main thread of an application. We will start by going over how multithreading should be handled in an Android application and the available technologies we now have to easily handle this. We will then move on to implement loading data from the internet using libraries such as Retrofit and OkHttp, after which we will look at how we can persist data on a device using libraries such as Room and DataStore.

In this chapter, we will cover the following main topics:

- Understanding Kotlin coroutines and flows
- Using OkHttp and Retrofit for networking
- Using the Room library for data persistence
- Understanding and using the DataStore library

By the end of this chapter, you will have become familiar with how we can load, manage, and persist data in an Android application.

Technical requirements

This chapter has the following hardware and software requirements:

- Android Studio Arctic Fox 2020.3.1 Patch 3

The code files for this chapter can be found here:

`https://github.com/PacktPublishing/Clean-Android-Architecture/tree/main/Chapter2`

Check out the following video to see the Code in Action: `https://bit.ly/38uecPi`

Understanding Kotlin coroutines and Flows

In this section, we will look at how threading works in the Android ecosystem and what applications must do to ensure that long-running operations do not block the user from using an application. We will then look at what available options we have available to execute operations in the background, with a focus on coroutines. Finally, we will look over Kotlin flows, which we can use to handle asynchronous work using a reactive and functional approach.

Android applications normally run in a single process on a user's device. When the operating system starts the application's process, it will allocate memory resources for the process to be executed. This process, when started, will have one thread of execution running within. This thread is referred to as the "main thread" or "**user interface (UI) thread**". In Android, this concept is very important because it is the thread that deals with user interaction. This imposes certain limitations for developers when dealing with it, as outlined here:

- The main thread must not be blocked by long-running or **input/output (I/O)** operations.

- All updates to the UI must be done on the main thread.

The idea is that the user should still be able to interact with an application as much as possible even if the application is doing some work. Every time we want to load and save data from or to the internet, local storage, content providers, and so on, we should use another thread or use multiple threads. The way the device's processor deals with multiple threads is by assigning a core for each thread. When there are more threads than cores, it will jump back and forth between every single instruction from each thread. Having too many threads being executed simultaneously will end up creating a bad **user experience (UX)** because the processor will now need to jump between the main thread and the rest of the threads being executed at the same time, so we will need to be mindful of how many threads are being executed concurrently.

In Java, a thread can be created using the `Thread` class; however, creating a new thread for every asynchronous operation is a very resource-expensive operation. Java also offers the concept of `ThreadPool` or `Executor`. These typically manage fixed a collection of threads that will be reused for different operations. Because of the Android restriction regarding updating the UI on the main thread, classes such as `Handler` and `Looper` were introduced, whereby you can submit the result of an operation performed on a background thread back on the main thread. An example of this is provided here:

```
class MyClass {

    fun asyncSum(a: Int, b: Int, callback: (Int) -> Unit) {
        val handler = Handler(Looper.getMainLooper())
        Thread(Runnable {
            val result = a + b
            handler.post(Runnable {
                callback(result)
            })
        }).start()
    }
}
```

In the preceding code snippet, the sum of two numbers will be performed on a new thread, and the result will then be posted back using the `Handler` object that is connected to the main `Looper` object, which itself will loop the main thread.

The repeated usage of `Handler` and `Looper` gave birth to `AsyncTask`, which offers the possibility of moving the necessary operations on a background thread and receiving the result on the main thread. `AsyncTask` worked with the same principle as the preceding example, only instead of creating a new thread for every new operation, it would by default use the same thread (although this later became configurable), which means that if two `AsyncTask` instances were executed at the same time, one would wait after the other. An example of the same sum operations might look like this:

```
    fun asyncSum(a: Int, b: Int, callback: (Int) -> Unit) {
        object : AsyncTask<Nothing, Nothing, Int>() {

            override fun doInBackground(vararg params:
                Nothing?): Int {
                return a+b
            }
```

```kotlin
            override fun onPostExecute(result: Int) {
                super.onPostExecute(result)
                callback(result)
            }

    }.execute()
}
```

In the preceding example, the sum is done in the `doInBackground` method, which is executed on a separate thread, and the `onPostExecute` method would be executed on the main thread.

Let's now imagine that we want to chain these sums and apply them multiple times, as follows:

```kotlin
fun asyncComplicatedSum(a: Int, b: Int, c: Int) {
    asyncSum(a, b) { tempSum ->
        asyncSum(tempSum, c) { finalSum ->
            Log.d(this.javaClass.name, "Final sum
                $finalSum")
        }
    }
}
```

In the preceding example, we try to sum two numbers and add the result to number `c`. As you can see, we need to use the callback and wait for `a` and `b` to finish and then apply the same function to the result of `a+b` and the number `c`.

Let's imagine what an application might look like when having to deal with loading data from multiple data sources, merging them together, handling errors, and stopping the asynchronous execution if the user leaves the current activity or fragment. The RxJava library tries to tackle all these problems through an event-driven approach. It introduces the concepts of streams and flows of data that can be observed, transformed, merged with other data streams, and executed on different threads. The sum of two numbers in RxJava might look something like this:

```kotlin
fun asyncSum(a: Int, b: Int): Single<Int> {
    return Single.create<Int> {
        it.onSuccess(a + b)
```

```
        }.subscribeOn(Schedulers.io())
            .observeOn(AndroidSchedulers.mainThread())
    }
```

In the preceding example, we create a `Single` instance, which is a stream that emits only one value (for emitting multiple values, we have the `Flowable` and `Observable` options). The value emitted is the sum of the two numbers. The usage of `subscribeOn` is for executing the upstream (the sum) on an I/O thread managed by RxJava internally, and the usage of `observeOn` is to have everything downstream (all the commands that will follow) to get the result on the main thread.

If we want to chain multiple sums, then we would have something like this:

```
fun asyncComplicatedSum(a: Int, b: Int, c: Int) {
        val disposable = asyncSum(a, b)
            .flatMap {
                asyncSum(it, c)
            }
            .subscribe ({
                Log.d(this.javaClass.name, "Final sum $it")
            },{
                Log.d(this.javaClass.name, "Something went
                    wrong")
            })
    }
```

In the preceding example, the sum of a and b is executed, then through the `flatMap` operator, we add c to that result. The usage of `subscribe` method is for triggering sums and listening for the results. This is because the `Single` instance used is a cold observable; it will only be executed only when `subscribe` is called. There is also the concept of hot observables, which will emit whether there are subscribers or not. The result of the `subscribe` operator will return a `Disposable` instance that offers a `dispose` method that can be called when we want to stop listening for data from the stream. This is useful in situations where our activities and fragments are destroyed, and we don't want to update our UI to avoid context leaks.

Kotlin coroutines

So far, we have analyzed technologies that revolve around the Java and Android frameworks. With the adoption of Kotlin, other technologies have emerged that deal with multithreading and are Kotlin-specific. One of these is the concept of coroutines. Coroutines simplify the way we write asynchronous code. Instead of dealing with callbacks, coroutines introduce the concept of scopes where we can specify which threads our blocks of code will execute in. The scopes can also connect to lifecycle-aware components that help us unsubscribe from the results of asynchronous work when our lifecycle-aware components terminate. Let's look at the following example of coroutines for the same sum:

```
suspend fun asyncSum(a: Int, b: Int): Int {
    return withContext(Dispatchers.IO) {
        a + b
    }
}
```

In the preceding example, the `withContext` method will execute the block of code inside it in the threads managed by the I/O dispatcher. The number of threads associated with this dispatcher is managed internally by the Kotlin framework and is associated with the number of cores the processor of the device has. This often means that we don't have to worry about the performance of our applications when multiple asynchronous operations are executed concurrently. Another interesting thing to note in the example is the usage of the `suspend` keyword. This is to alert the caller of this method that it will be executed using coroutines on a separate thread.

Now, let's see what things will look like when we want to invoke this method. Have a look at the following code snippet:

```
class MyClass : CoroutineScope {
    override val coroutineContext: CoroutineContext
        get() = Dispatchers.Main + job

    private lateinit var job: Job

    fun asyncComplicatedSum(a: Int, b: Int, c: Int) {
        launch {
            try {
                val tempSum = asyncSum(a, b)
                val finalSum = asyncSum(tempSum, c)
```

```
                    Log.d(this.javaClass.name, "Final sum
                        $finalSum")
                } catch (e: Exception) {
                    Log.d(this.javaClass.name, "Something went
                        wrong")
                }
            }
        }
    }

    fun create() {
        job = Job()
    }

    fun destroy() {
        job.cancel()
    }
}
```

In `asyncComplicatedSum`, we use the `launch` method. This method is associated with the `CoroutineContext` object defined in this class. The context is defined using the `Main` dispatcher combined with the `Job` object that will be associated with the lifecycle of this object. If the `destroy` method is called while we are waiting for the result of the sum, then the execution of the sum will stop and we will stop getting the result of the sum. The code will execute each of the sums on the I/O thread and then execute log statements on the main thread if the job is still alive.

In Android, we already have a few `CoroutineScope` objects already defined and associated with our lifecycle-aware classes. One that will be relevant to us is the one defined for `ViewModels`. This can be found in the `org.jetbrains.kotlinx:kotlinx-coroutines-android` library and will look something like this:

```
class MyViewModel: ViewModel() {
    init {
        viewModelScope.launch {  }
    }
}
```

`viewModelScope` is a Kotlin extension created for `ViewModel` instances that will execute if the `ViewModel` instance is alive. If `onCleared` is called on the `ViewModel` instance, then it will stop listening to the remaining code to be executed in the `launch` block.

In this section, we've analyzed how Kotlin coroutines work and how we can use them to handle asynchronous operations in an Android application. In the next section, we will create an Android application that will use Kotlin coroutines for a simple asynchronous operation.

Exercise 02.01 – Using Kotlin coroutines

Create an application that will display two input fields, one text field, and a button. The input fields will be limited to numbers only, and when the user presses the button, then the text field will display the sum of the two numbers after 5 seconds. The sum and waiting will be implemented using coroutines.

To complete the exercise, you will need to build the following:

- A class that will perform the addition of the two numbers

- A `ViewModel` class that will invoke the addition

- The UI using Compose that will use the following function:

```
@Composable
fun Calculator(
    a: String,
    onAChanged: (String) -> Unit,
    b: String,
    onBChanged: (String) -> Unit,
    result: String,
    onButtonClick: () -> Unit
) {
    Column(modifier = Modifier.padding(16.dp)) {
        OutlinedTextField(
            value = a,
            onValueChange = onAChanged,
            keyboardOptions = KeyboardOptions
                (keyboardType = KeyboardType.Number),
            label = { Text("a") }
        )
```

```
        OutlinedTextField(
            value = b,
            onValueChange = onBChanged,
            keyboardOptions = KeyboardOptions
                (keyboardType = KeyboardType.Number),
            label = { Text("b") }
        )
        Text(text = result)
        Button(onClick = onButtonClick) {
            Text(text = "Calculate")
        }
    }
}
```

Follow these steps to complete the exercise:

1. Create a new project in Android Studio using an **Empty Compose Activity**.

2. At the top level of the `build.gradle` file, define the Compose library version as follows:

```
buildscript {
    ext {
        compose_version = '1.0.5'
    }
    ...
}
```

3. In the `app/build.gradle` file, we need to add the following dependencies:

```
dependencies {
    implementation 'androidx.core:core-ktx:1.7.0'
    implementation 'androidx.appcompat:appcompat:1.4.0'
    implementation 'com.google.android.
material:material:1.4.0'
    implementation "androidx.compose.ui:ui:$compose_
version"
    implementation "androidx.compose.
material:material:$compose_version"
    implementation "androidx.compose.ui:ui-tooling-
```

```
preview:$compose_version"
    implementation 'androidx.lifecycle:lifecycle-runtime-
ktx:2.4.0'
    implementation 'androidx.activity:activity-
compose:1.4.0'
    implementation 'org.jetbrains.kotlinx:kotlinx-
coroutines-android:1.5.0'
    implementation "androidx.lifecycle:lifecycle-
viewmodel-ktx:2.4.0"
    implementation "androidx.lifecycle:lifecycle-
viewmodel-compose:2.4.0"
    testImplementation 'junit:junit:4.13.2'
    androidTestImplementation 'androidx.test.
ext:junit:1.1.3'
    androidTestImplementation 'androidx.test.
espresso:espresso-core:3.4.0'
    androidTestImplementation "androidx.compose.ui:ui-
test-junit4:$compose_version"
    testImplementation "org.jetbrains.kotlinx:kotlinx-
coroutines-test:1.5.0"
    debugImplementation "androidx.compose.ui:ui-
tooling:$compose_version"
}
```

4. Start by creating a `NumberAdder` class and define an `add` operation and a delay, as follows:

```
private const val DELAY = 5000
class NumberAdder(
    private val dispatcher: CoroutineDispatcher =
        Dispatchers.IO,
    private val delay: Int = DELAY
) {

    suspend fun add(a: Int, b: Int): Int {
        return withContext(dispatcher) {
            delay(delay.toLong())
            a + b
```

```
            }
        }
    }
```

In this class, we will add our 5-second delay before performing the sum of the two numbers. This is to highlight the asynchronous operation more. `CoroutineDispatcher` and the amount we want to delay by will be injected through the constructor. This is because we want to unit-test this class.

5. Next, we will need to unit-test this class. Before we write the test, create a test rule so that we can reuse it for coroutines, as follows:

```
class DispatcherTestRule : TestRule {

    @ExperimentalCoroutinesApi
    val testDispatcher = TestCoroutineDispatcher()

    @ExperimentalCoroutinesApi
    override fun apply(base: Statement?, description:
        Description?): Statement {
        try {
            Dispatchers.setMain(testDispatcher)
            base?.evaluate()
        } catch (e: Exception) {

        } finally {
            Dispatchers.resetMain()
            testDispatcher.cleanupTestCoroutines()
        }
        return base!!
    }
}
```

In this class, we create a `TestCoroutineDispatcher` instance that will later be injected into the unit test so that the test can execute the sum in a synchronous way. `@ExperimentalCoroutinesApi` suggests that the usage of `TestCoroutineDispatcher` is still in an experimental state and will be moved to a stable version in the future.

6. Now, write the unit test for the class, in the form of `NumberAdderTest`, as follows:

```
class NumberAdderTest {

    @get:Rule
    val dispatcherTestRule = DispatcherTestRule()

    @ExperimentalCoroutinesApi
    @Test
    fun testAdd() = runBlockingTest {
        val adder = NumberAdder(dispatcherTestRule.
            testDispatcher, 0)
        assertEquals(5, adder.add(1, 4))

    }
}
```

Here, we inject the `testDispatcher` object we created in `DispatcherTestRule` into `NumberAdder`, and we then invoke the add function. The entire test is executed in a special `CoroutineScope` block called `runBlockingTest`, that will ensure all the coroutines launched must complete.

7. Next, go ahead and create a `ViewModel` class, like this:

```
class MainViewModel(private val adder: NumberAdder =
NumberAdder()) : ViewModel() {

    var resultState by mutableStateOf("0")
        private set

    fun add(a: String, b: String) {
        viewModelScope.launch {
            val result = adder.add(a.toInt(),
                b.toInt())
            resultState = result.toString()
        }
    }
}
```

Here, we use a Compose state that will retain the result of the addition, and a method that will trigger the addition into `viewModelScope`.

8. After the `ViewModel` class has been created, go ahead and create an activity class, as follows:

```
class MainActivity : ComponentActivity() {
    override fun onCreate(savedInstanceState: Bundle?) {
        super.onCreate(savedInstanceState)
        setContent {
            Exercise201Theme {
                Surface {
                    Screen()
                }
            }
        }
    }
}
```

Here, we initialize our activity with the content. `Exercise201Theme` should be replaced with the theme generated by Android Studio when the project was created. Typically, this should be in a `Theme` file and should be a `@Composable` function that has the application name followed by the `Theme` suffix. If that is not available, you can use `MaterialTheme` instead for the purpose of the exercise.

9. Next, create a `Screen` function, as follows:

```
@Composable
fun Screen(viewModel: MainViewModel = viewModel()) {
    var a by remember { mutableStateOf("") }
    var b by remember { mutableStateOf("") }
    Calculator(
        a = a,
        onAChanged = {
            a = it
        },
        b = b,
        onBChanged = {
            b = it
        },
```

```
        result = viewModel.resultState,
        onButtonClick = {
            viewModel.add(a, b)
        })
    }
```

In this method, we define variables for our text fields, then we pass the result of the addition of the numbers from the ViewModel, and finally, we invoke the ViewModel to perform the addition.

10. And finally, add the `Calculator` function from the exercise definition to the `MainActivity` file.

If we run the preceding example, we should see our UI elements, and after inserting the numbers and clicking the button, we will get our result. One thing to notice is that the user will be able to interact with the UI while the `add` method is executed, and clicking multiple times for different numbers will get the results 5 seconds after each button press.

Using coroutines can improve the quality of an Android application, especially when combined with Android extensions for the `ViewModel` class and lifecycle-aware components. Coroutines simplify the code we write for asynchronous operations, and the addition of the `suspend` keyword can enforce more rigor when dealing with these operations.

Kotlin Flows

Coroutines offer a good solution for dealing with asynchronous operations; however, they do not offer a good ability to handle multiple streams of data in the same way RxJava does. Flows represent an extension to coroutines, which is meant to solve this problem. When dealing with flows, there are three entities to consider, as outlined here:

- **Producer**: This entity is responsible for emitting the data.

- **Intermediary**: This entity deals with the transformation or manipulation of the data.

- **Consumer**: This entity consumes the data in the stream.

Let's look at the following example of adding two numbers and how it might look like using Kotlin flows:

```
fun asyncSum(a: Int, b: Int): Flow<Int> {
        return flow {
            this.emit(a + b)
```

```
        }.flowOn(Dispatchers.IO)
    }
```

Here, we create a `Flow` object that will emit the result of a + b on a stream. The `flowOn` method will move the execution of the upstream on an I/O thread. Here, we note the similarity to RxJava in the concept of how `Flows` work, but we also notice that it's an extension of coroutines because of the use of `Dispatchers`. Let's now look at how flows look on the consumer side, as follows:

```
class MyClass : CoroutineScope {
    override val coroutineContext: CoroutineContext
        get() = Dispatchers.Main + job

    private lateinit var job: Job

    @FlowPreview
    fun asyncComplicatedSum(a: Int, b: Int, c: Int) {
        launch {
            asyncSum(a, b)
                .flatMapConcat {
                    asyncSum(it, c)
                }
                .catch {
                    Log.d(this.javaClass.name, "Something
                        went wrong")
                }
                .collect {
                    Log.d(this.javaClass.name, "Final sum
                        $it")
                }
        }
    }
}
```

Here, we also notice similarities to RxJava—that is, when we try to manipulate the stream to perform the addition to number c and when it comes to error handling due to the `catch` method. The `collect` method, however, is closer to coroutines, and it requires a `CoroutineScope` to be used or to declare the calling method as a suspend one.

Flows offer a couple of specialized classes for particular use cases: `StateFlow` and `SharedFlow`. The `StateFlow` class is useful because it will offer subscribers the last value stored when they subscribe, like how `LiveData` works. Flows can also be cold and hot, and `SharedFlow` is a specialized implementation of a hot flow. `SharedFlow` will emit items if it is kept in memory if there are any consumers or not. When a consumer subscribes to `SharedFlow`, it will also emit the last value stored to the consumer, as with `StateFlow`.

In this section, we have looked at Kotlin flows and the benefits they provide when it comes to handling asynchronous operations. Next, we will look at how we can use Kotlin flows in an Android application through a simple exercise.

Exercise 02.02 – Using Kotlin Flows

Modify the application from *Exercise 02.01* so that the addition of the two numbers will return a Flow instead of a suspended function.

To complete the exercise, you will need to do the following:

- Rewrite the add function in `NumberAdder` to return a Flow.

- Change how the `ViewModel` invokes the add function.

Follow these steps to complete the exercise:

1. Modify the add function in `NumberAdder` to return a Flow, as follows:

```
private const val DELAY = 5000
class NumberAdder(
    private val dispatcher: CoroutineDispatcher =
        Dispatchers.IO,
    private val delay: Int = DELAY
) {

    suspend fun add(a: Int, b: Int): Flow<Int> {
        return flow {
            emit(a + b)
        }.onEach {
            delay(delay.toLong())
        }.flowOn(dispatcher)
    }
}
```

Here, we create a new Flow where we emit the sum of a and b, after which we put a delay on each item emitted in the stream, and finally, we specify the CoroutineDispatcher instance we wish to execute the sum on.

2. Next, let's modify the unit test for the sum, as follows:

```
class NumberAdderTest {

    @get:Rule
    val dispatcherTestRule = DispatcherTestRule()

    @ExperimentalCoroutinesApi
    @Test
    fun testAdd() = runBlockingTest {
        val adder = NumberAdder
            (dispatcherTestRule.testDispatcher, 0)
        val result = adder.add(1, 4).first()
        assertEquals(5, result)
    }
}
```

Because the add method returns a Flow object, we must now find the first item emitted in the flow and assert the value of that item against our expected result.

3. Modify the MainViewModel class to consume the add operation, as follows:

```
class MainViewModel(private val adder: NumberAdder =
NumberAdder()) : ViewModel() {

    var resultState by mutableStateOf("0")
        private set

    fun add(a: String, b: String) {
        viewModelScope.launch {
            adder.add(a.toInt(), b.toInt())
                .collect {
                    resultState = it.toString()
                }
        }
    }
}
```

Here, the add method will still use the same `CoroutineScope` instance to launch the add method, which will now use the `collect` method to get the result of the sum.

If we launch the application after following the steps from the exercise, the behavior will be the same as for *Exercise 02.01*, and we can see how Kotlin flows extend the functionality of coroutines by introducing concepts from RxJava to simplify how we can handle multiple streams of data.

In this section, we've seen how handling asynchronous operations has evolved over time and how much our applications benefit from concepts such as coroutines and flows that provide management for background threads, simplify how we execute asynchronous operations, manage multiple streams of data, and can connect to the lifecycle of Android components. In the following section, we will look at tools we can use to fetch data from the network and how they can be integrated with Kotlin coroutines and flows.

Using OkHttp and Retrofit for networking

In this section, we will look at how we can use the Retrofit library to perform networking operations and the benefits it provides.

Many Android applications require the internet to access data stored on various servers. Often, this is done through the **HyperText Transfer Protocol** (**HTTP**) protocol in which data is exchanged between the applications and the servers. The data is often represented in **JavaScript Object Notation** (**JSON**) format. In the past, these types of exchanges were implemented either with `HttpURLConnection` or Apache HttpClient. Working with either of these components meant that developers would need to manually handle the conversion from **plain old Java objects** (**POJOs**) to JSON, handle various network configurations, and deal with backward compatibility.

The OkHttp library will address some of these issues through an `OkHttpClient` class that will handle various network configurations and that provides other features such as caching. The Retrofit library, which can be placed on top of the OkHttp library, is meant to ensure type safety when dealing with various data formats. It's very configurable and allows the possibility to plug in various converter libraries for POJO-to-JSON conversion or **Extensible Markup Language** (**XML**) or other types of formats.

In order to add Retrofit and OkHttp to the project, we will add the following dependencies to the `build.gradle` file:

```
dependencies {
    …
    implementation "com.squareup.okhttp3:okhttp:4.9.0"
```

```
    implementation "com.squareup.retrofit2:retrofit:2.9.0"
    ...
}
```

Next, we will need to determine which converters we will need to use for the data. Because JSON is a common format, we will use a JSON converter and the Moshi library to do so, so we will need to add dependencies to these two libraries, as follows:

```
dependencies {
    ...
    implementation "com.squareup.okhttp3:okhttp:4.9.0"
    implementation "com.squareup.retrofit2:retrofit:2.9.0"
    implementation "com.squareup.retrofit2:converter-
moshi:2.9.0"
    implementation "com.squareup.moshi:moshi:1.13.0"
    ...
}
```

Here, the Moshi library will be responsible for converting POJOs into JSON, and the converter library will plug into the Retrofit library and trigger this conversion when data is exchanged between the Android application and the server.

Let's assume we will need to fetch data from a server in a JSON format. We can use the `https://jsonplaceholder.typicode.com/` service as an example. If we want to fetch a list of users, we can use the `https://jsonplaceholder.typicode.com/users` **Uniform Resource Locator** (**URL**). A user's JSON representation looks like this:

```
{
"id": 1,
    "name": "Leanne Graham",
    "username": "Bret",
    "email": "Sincere@april.biz",
    "address": {
      "street": "Kulas Light",
      "suite": "Apt. 556",
      "city": "Gwenborough",
      "zipcode": "92998-3874",
      "geo": {
        "lat": "-37.3159",
```

```
        "lng": "81.1496"
    }
},
"phone": "1-770-736-8031 x56442",
"website": "hildegard.org",
"company": {
  "name": "Romaguera-Crona",
  "catchPhrase": "Multi-layered client-server neural-
     net",
  "bs": "harness real-time e-markets"
}
```

We can see in the JSON representation that the user has an id, a username, an
email value, and so on. In Kotlin, we can create a representation of this, and we can
exclude properties that the application doesn't need, such as email, address, phone,
website, and company, as follows:

```
data class User(
    @Json(name = "id") val id: Long,
    @Json(name = "name") val name: String,
    @Json(name = "username") val username: String
)
```

Here, we are using Moshi to map the property from a JSON to a Kotlin type, and we only
kept three of the fields present in the initial JSON. Now, let's look at how we can initialize
our networking libraries. The code to accomplish this is shown in the following snippet:

```
fun createOkHttpClient() =  OkHttpClient
    .Builder()
    .readTimeout(15, TimeUnit.SECONDS)
    .connectTimeout(15, TimeUnit.SECONDS)
    .build()
```

For OkHttp, we use a Builder method to create a new OkHttpClient instance,
and we can provide certain configurations for it. We will now use the OkHttpClient
instance created previously to create a Retrofit instance, as follows:

```
fun createRetrofit(
    okHttpClient: OkHttpClient
): Retrofit {
```

```
        return Retrofit.Builder()
            .baseUrl("https://jsonplaceholder.typicode.com/")
            .client(okHttpClient)
            .build()
    }
```

Here, we create a new `Retrofit` instance that will have the base URL set to `https://jsonplaceholder.typicode.com/`. Changing the base URL comes in handy during development. Many teams will have a development URL used internally to test the development and integration of features and will have a production URL where the actual user data is set. Now, we will need to connect the Moshi JSON serialization to the `Retrofit` instance, as follows:

```
Fun createConverterFactory(): MoshiConverterFactory =
MoshiConverterFactory.create()
```

Here, we create `MoshiConverterFactory`, which is a Retrofit converter designed to connect `Retrofit` to the JSON serialization done by Moshi. We will now need to change our `Retrofit` initialization to what follows:

```
fun createRetrofit(
        okHttpClient: OkHttpClient,
        gsonConverterFactory: MoshiConverterFactory
    ): Retrofit {
        return Retrofit.Builder()
 .baseUrl("https://jsonplaceholder.typicode.com/")
            .client(okHttpClient)
            .addConverterFactory(gsonConverterFactory)
            .build()
    }
```

Here, we add the `MoshiConverterFactory` converter to the Retrofit `Builder` method to allow the two components to work together. Finally, we can create a Retrofit interface that will have templates for the HTTP request, as follows:

```
interface UserService {

        @GET("/users")
        fun getUsers(): Call<List<User>>
```

```
    @GET("/users/{userId}")
    fun getUser(@Path("userId") userId: Int):
        Call<User>


    @POST("/users")
    fun createUser(@Body user: User): Call<User>


    @PUT("/users/{userId}")
    fun updateUser(@Path("userId") userId: Int, @Body
        user: User): Call<User>
}
```

This interface contains an example of various methods for getting, creating, updating, and deleting data on servers. Note that the return type of these methods is a `Call` object that offers the ability to execute HTTP requests synchronously or asynchronously. One of the things that makes Retrofit more appealing to developers is the fact that it can be integrated with other asynchronous libraries such as RxJava and coroutines. Translating the preceding example to coroutines will look something like this:

```
interface UserService {

    @GET("/users")
    suspend fun getUsers(): List<User>


    @GET("/users/{userId}")
    suspend fun getUser(@Path("userId") userId: Int):
        User


    @POST("/users")
    suspend fun createUser(@Body user: User): User


    @PUT("/users/{userId}")
    suspend fun updateUser(@Path("userId") userId: Int,
        @Body user: User): User
}
```

In the preceding example, we add the `suspend` keyword to each method and we remove the dependency to the `Call` class. This allows us to execute these methods using coroutines. To create an instance of this class, we need to do the following:

```
fun createUserService(retrofit: Retrofit) = retrofit.
create(UserService::class.java)
```

Here, we use the `Retrofit` instance created previously to create a new instance of `UserService`.

In this section, we have analyzed how we can use OkHttp and Retrofit to load data from the internet and the benefits these libraries provide, especially when combined with Kotlin coroutines and flows. In the next section, we will create an Android application that will use these libraries to fetch and display data on the UI.

Exercise 02.03 – Using OkHttp and Retrofit

Create an Android application that connects to `https://jsonplaceholder.typicode.com/` and displays a list of users using OkHttp, Retrofit, and Moshi. For each user, we will display the name, username, and email.

To complete the exercise, you will need to do the following:

- Create a `User` data class that will map the JSON representation of the user.
- Create a `UserService` class that will have a method to retrieve a list of users.
- Create a `ViewModel` class that will use `UserService` to retrieve a list of users.
- Implement an `Activity` class that will display a list of users.

A UI list will be created using the following method:

```
@Composable
fun UserList(users: List<User>) {
    LazyColumn(modifier = Modifier.padding(16.dp)) {
        items(users) {
            Column(modifier = Modifier.padding(16.dp)) {
                Text(text = it.name)
                Text(text = it.username)
                Text(text = it.email)
            }
        }
    }
}
```

```
        }
    }
```

Follow these steps to complete the exercise:

1. Create an Android application with an **Empty Compose Activity**.

2. At the top level of the `build.gradle` file, define the Compose library version, as follows:

```
buildscript {
    ext {
        compose_version = '1.0.5'
    }
    ...
}
```

3. In the `app/build.gradle` file, add the following dependencies:

```
dependencies {
    implementation 'androidx.core:core-ktx:1.7.0'
    implementation 'androidx.appcompat:appcompat:1.4.0'
    implementation 'com.google.android.
material:material:1.4.0'
    implementation "androidx.compose.ui:ui:$compose_
version"
    implementation "androidx.compose.
material:material:$compose_version"
    implementation "androidx.compose.ui:ui-tooling-
preview:$compose_version"
    implementation 'androidx.lifecycle:lifecycle-runtime-
ktx:2.4.0'
    implementation 'androidx.activity:activity-
compose:1.4.0'
    implementation 'org.jetbrains.kotlinx:kotlinx-
coroutines-android:1.5.0'
    implementation "androidx.lifecycle:lifecycle-
viewmodel-ktx:2.4.0"
    implementation "androidx.lifecycle:lifecycle-
viewmodel-compose:2.4.0"
    implementation "com.squareup.okhttp3:okhttp:4.9.0"
```

```
    implementation "com.squareup.retrofit2:retrofit:2.9.0"
    implementation "com.squareup.retrofit2:converter-
moshi:2.9.0"
    implementation "com.squareup.moshi:moshi:1.13.0"
    implementation "com.squareup.moshi:moshi-
kotlin:1.13.0"
    testImplementation 'junit:junit:4.13.2'
    androidTestImplementation 'androidx.test.
ext:junit:1.1.3'
    androidTestImplementation 'androidx.test.
espresso:espresso-core:3.4.0'
    androidTestImplementation "androidx.compose.ui:ui-
test-junit4:$compose_version"
    testImplementation "org.jetbrains.kotlinx:kotlinx-
coroutines-test:1.5.0"
    debugImplementation "androidx.compose.ui:ui-
tooling:$compose_version"
}
```

4. Now, add a permission for internet access to the `AndroidManifest.xml` file, as follows:

    ```
    <uses-permission android:name="android.permission.
    INTERNET"/>
    ```

5. Now move on and create a class that will hold the user information, as follows:

    ```
    @JsonClass(generateAdapter = true)
    data class User(
        @Json(name = "id") val id: Long,
        @Json(name = "name") val name: String,
        @Json(name = "username") val username: String,
        @Json(name = "email") val email: String
    )
    ```

    Here, we will hold the `id` field, which is generally a relevant field for distinguishing between different users and fields that we are required to display.

6. Next, create a `UserService` class that will fetch the user data, as follows:

    ```
    interface UserService {
    ```

```
@GET("/users")
suspend fun getUsers(): List<User>


}
```

Here, we will only have one method that will get a list of users from the /users path.

7. Now, we initialize the networking objects. Because we aren't using any **dependency injection (DI)** frameworks and we only need to create one instance of each, we will hold the objects in the MainApplication class, as follows:

```
class MyApplication : Application() {


    companion object {
        lateinit var userService: UserService
    }


    override fun onCreate() {
        super.onCreate()
        val okHttpClient = OkHttpClient
            .Builder()
            .readTimeout(15, TimeUnit.SECONDS)
            .connectTimeout(15, TimeUnit.SECONDS)
            .build()
        val moshi = Moshi.Builder().
            add(KotlinJsonAdapterFactory()).build()
        val retrofit = Retrofit.Builder()
            .baseUrl("https://jsonplaceholder.typicode.
com/")
            .client(okHttpClient)
            .addConverterFactory(MoshiConverterFactory.
create(moshi))
            .build()
        userService = retrofit.create(UserService::class.
java)
    }
}
```

Here, we are initializing our networking libraries and the `UserService` object. Currently, we are holding a static reference to this object, which is not a good idea in general. Normally, we would rely on DI frameworks to manage these networking dependencies.

8. In the `AndroidManifest.xml` file, add the following code:

```
<application
    ...
    android:name=".MyApplication"
    ...>
```

Given that we are inheriting from the `Application` class, we will need to add this class to the manifest.

9. Next, go ahead and create a `MainViewModel` class, as follows:

```
class MainViewModel(private val userService:
    UserService) : ViewModel() {

    var resultState by mutableStateOf
        <List<User>>(emptyList())
        private set

    init {
        viewModelScope.launch {
            val users = userService.getUsers()
            resultState = users
        }
    }
}

class MainViewModelFactory : ViewModelProvider.Factory {
    override fun <T : ViewModel> create(modelClass:
        Class<T>): T =
        MainViewModel(MyApplication.userService) as T
}
```

The `MainViewModel` class will depend on the `UserService` class to get a list of `Users` and store them in a Compose state that will be used in the UI. Here, we are also creating a `MainViewModelFactory` class that will be responsible for injecting the `UserService` class into the `MainViewModel` class.

10. Now, we move on and create a `MainActivity` class, as follows:

```
class MainActivity : ComponentActivity() {
    override fun onCreate(savedInstanceState: Bundle?) {
        super.onCreate(savedInstanceState)
        setContent {
            Exercise0203Theme {
                Surface {
                    Screen()
                }
            }
        }
    }
}
```

Here, we initialize our activity with the content. The `Exercise203Theme` theme should be replaced with the theme generated by Android Studio when the project was created. Typically, this should be in a `Theme` file and should be a `@Composable` function that has the application name followed by the `Theme` suffix. If that is not available, you can use `MaterialTheme` instead for the purpose of the exercise.

11. Create a `Screen` method in which we will grab a list of users from the `MainViewModel` class and draw a list of items, as follows:

```
@Composable
fun Screen(viewModel: MainViewModel = viewModel
    (factory = MainViewModelFactory())) {
    UserList(users = viewModel.resultState)
}
```

12. And finally, add the `UserList` function from the exercise definition into the `MainActivity` file.

If we launch the application after following the steps from the exercise, we should be able to see a list of users being loaded if the device has internet access.

In this section, we have seen how we can typically retrieve data from the internet in an Android application. We have looked at libraries such as OkHttp and Retrofit and seen how straightforward it is to make HTTP calls in a type-safe way without converting JSON files to data classes manually. We have also observed the potential of these libraries due to their integration with asynchronous technologies such as RxJava and coroutines. In the following section, we will look at libraries used for persisting data and how we can integrate them with networking libraries as well as coroutines and flows.

Using the Room library for data persistence

In this section, we will discuss how to persist data in Android applications and how we can use the Room library to do this.

Android offers many ways for persisting data on an Android device, mostly involving files. Some of these files have a specialized approach to persisting data. One of these approaches is in the form of SQLite. SQLite is a special type of file in which structured data can be stored using **Structured Query Language** (**SQL**) queries, as with other types of databases such as MySQL and Oracle.

In the past, if developers wanted to persist data in SQLite, they were required to manually define tables, write queries, and transform objects containing this data into the appropriate formats for performing **create, read, update, and delete** (**CRUD**) operations. This type of work involved a load of boilerplate code that was susceptible to bugs. Room is the answer to that by providing an abstraction layer on top of the SQLite operations.

In order to add Room to an application, we will need to add the following libraries in `build.gradle`:

```
dependencies {
    ...
    implementation "androidx.room:room-runtime:2.4.0"
    kapt "androidx.room:room-compiler:2.4.0"
    ...
}
```

The reason for the `kapt` usage is that Room uses annotations that will generate the code required for the interaction with the SQLite layer. In order to use the `kapt` feature, we will need to add the plugin to the `build.gradle` file, as follows:

```
plugins {
    ...
```

```
    id 'kotlin-kapt'
}
```

This will allow the build system to analyze annotations across the project that require code generation and generate the necessary classes based on the provided annotations.

The data we want to store is annotated with the `@Entity` annotation, as illustrated in the following code snippet:

```
@Entity(tableName = "user")class UserEntity(
    @PrimaryKey @ColumnInfo(name = "id") val id: Long,
    @ColumnInfo(name = "name") val name: String,
    @ColumnInfo(name = "username") val username: String
)
```

Here, we have defined a Room entity named `UserEntity` that will represent a table named `user` and has the **primary key (PK)** set to be the **identifier (ID)** of the user. The `@ColumnInfo` annotation is for the name the column will have in the database.

A typical set of CRUD operations might look like this:

```
@Dao
interface UserDao {
    @Query("SELECT * FROM user")
    fun getAll(): List<UserEntity>

    @Query("SELECT * FROM user WHERE id IN (:userIds)")
    fun loadAllByIds(userIds: IntArray): List<UserEntity>

    @Insert
    fun insert(vararg users: User)

    @Update
    fun update(vararg users: User)

    @Delete
    fun delete(user: User)
}
```

Just as how we defined in Retrofit a service interface to communicate with the server, we also define a similar interface for Room that we annotate with @Dao, for **data access object (DAO)**. In this example, we have defined a set of functions for getting all users stored in a table, finding users, inserting new users, updating a user, and deleting a user.

As with Retrofit, Room also provides integrations with coroutines, as illustrated in the following code snippet:

```
@Dao
interface UserDao {
    @Query("SELECT * FROM user")
    suspend fun getAll(): List<UserEntity>

    @Query("SELECT * FROM user WHERE id IN (:userIds)")
    suspend fun loadAllByIds(userIds: IntArray):
        List<UserEntity>

    @Insert
    suspend fun insert(vararg users: User)

    @Update
    suspend fun update(vararg users: User)

    @Delete
    suspend fun delete(user: User)
}
```

In the preceding example, we add the `suspend` keyword, which makes the Room library easy to integrate and execute as part of a coroutine.

On top of coroutines, the Room library also can integrate with Kotlin flows. This is useful for queries that will emit events every time a particular table has changed. This integration will look something like this:

```
@Dao
interface UserDao {
    @Query("SELECT * FROM user")
    fun getAll(): Flow<List<UserEntity>>
```

```
    @Query("SELECT * FROM user WHERE id IN (:userIds)")
    fun loadAllByIds(userIds: IntArray):
        Flow<List<UserEntity>>
}
```

In the preceding example, we have changed the @Query functions to return a Flow object. If a change occurs in the user table, then the queries will be re-triggered and a new list of users will be emitted.

We will now need to set up the database, as follows:

```
@Database(entities = [UserEntity::class], version = 1)
abstract class AppDatabase : RoomDatabase() {
    abstract fun userDao(): UserDao
}
```

In the preceding snippet, we define a new class that extends from the RoomDatabase class and use the @Database annotation to declare our entities and the current version. This version is used to keep track of migrations when the structure of the database changes in between new releases of our application.

To initialize the database, we will need to execute the following code:

```
val db = Room.databaseBuilder(
            applicationContext,
            AppDatabase::class.java, "name"
        ).build()
```

This will create our SQLite database and will return an instance of AppDatabase where we can access the DAO objects we have defined and invoke their methods to process the data.

In this section, we have looked at how we can persist data using Room and how it can be integrated with coroutines and flows. In the next section, we will create an Android application that will use Room to persist data and look at how it can be integrated with Retrofit and OkHttp.

Exercise 02.04 – Using Room to persist data

Integrate Room into *Exercise 02.03* so that when the users are loaded from Retrofit, they will be stored in the database and then displayed on the UI.

To complete the exercise, you will need to do the following:

1. Create a `UserEntity` class that will be a Room entity.

2. Create a `UserDao` class that will contain methods for inserting users and querying all the users as flows.

3. Create an `AppDatabase` class that will represent the application's database.

4. Modify the `MainViewModel` class to fetch users from the `UserService` class and then insert them into the `UserDao` class.

5. Modify the `MainActivity` class to use a list of `UserEntity` objects instead of `User` objects.

Follow these steps to complete the exercise:

1. Add the `kapt` plugin to the `app/build.gradle` file, as follows:

    ```
    plugins {
        ...
        id 'kotlin-kapt'
    }
    ```

2. Add Room dependencies to `app/build.gradle`, as follows:

    ```
    dependencies {
        ...
        implementation "androidx.room:room-runtime:2.4.0"
        implementation "androidx.room:room-ktx:2.4.0"
        kapt "androidx.room:room-compiler:2.4.0"
        ...
    }
    ```

3. Create a `UserEntity` class, as follows:

    ```
    @Entity(tableName = "user")
    class UserEntity(
        @PrimaryKey @ColumnInfo(name = "id") val id: Long,
        @ColumnInfo(name = "name") val name: String,
        @ColumnInfo(name = "username") val username:
            String,
        @ColumnInfo(name = "email") val email: String
    )
    ```

The `UserEntity` class has the same fields as the `User` class, and it contains the Room annotations for the table name and the names of each column.

4. Next, create a `UserDao` class, as follows:

```
@Dao
interface UserDao {

    @Query("SELECT * FROM user")
    fun getUsers(): Flow<List<UserEntity>>

    @Insert(onConflict = OnConflictStrategy.REPLACE)
    fun insertUsers(users: List<UserEntity>)
}
```

Here, we are using flows to return a list of users, and we use the `OnConflictStrategy.REPLACE` option so that if the same user is inserted multiple times, then it will be replaced with the one that will be inserted. Other options include `OnConflictStrategy.ABORT`, which will drop the entire transaction if a conflict occurs, or `OnConflictStrategy.IGNORE`, which will skip inserting rows where a conflict occurs.

5. Now, go ahead and create an `AppDatabase` class, as follows:

```
@Database(entities = [UserEntity::class], version = 1)
abstract class AppDatabase : RoomDatabase() {
    abstract fun userDao(): UserDao
}
```

In `AppDatabase`, we provide the `UserDao` class to be accessed and we use the `UserEntity` class for the users' table.

6. Next, we will need to initialize the `AppDatabase` object, as follows:

```
class MyApplication : Application() {

    companion object {

        ...

        lateinit var userDao: UserDao

        ...
    }
```

```
    override fun onCreate() {
        super.onCreate()
        ...
        val db = Room.databaseBuilder(
            applicationContext,
            AppDatabase::class.java, "my-database"
        ).build()
        userDao = db.userDao()
        ...
    }
}
```

Here, we are having the same issues that we had for Retrofit, so we will follow the same approach and use the Application class. Just as with Retrofit, a DI framework will help us solve this problem.

7. Now, let's integrate Room into the MainViewModel class, as follows:

```
class MainViewModel(
    private val userService: UserService,
    private val userDao: UserDao
) : ViewModel() {

    var resultState by
    mutableStateOf<List<UserEntity>>(emptyList())
        private set

    init {
        viewModelScope.launch {
            flow { emit(userService.getUsers()) }
                .onEach {
                    val userEntities =
                        it.map { user -> UserEntity
                            (user.id, user.name,
                                user.username, user.email) }
                    userDao.insertUsers(userEntities)
                }.flatMapConcat { userDao.getUsers() }
                .catch { emitAll(userDao.getUsers()) }
```

```
                        .flowOn(Dispatchers.IO)
                    .collect {
                        resultState = it
                    }
            }
        }
    }
    class MainViewModelFactory : ViewModelProvider.Factory {
        override fun <T : ViewModel> create(modelClass:
    Class<T>): T =
            MainViewModel(MyApplication.userService,
    MyApplication.userDao) as T
    }
```

The MainViewModel class now has a new dependency on the UserDao class. In the init block, we now create a flow in which we emit a list of users obtained from Retrofit that is then converted into UserEntity and inserted into the database. After this, we will query the UserEntities instances and return them in a stream that will be the result. If we have an error, we will return the current stored users.

8. Finally, update the type of users in the MainActivity class, as follows:

```
    class MainActivity : ComponentActivity() {
    ...
    @Composable
    fun UserList(users: List<UserEntity>) {
    ...
        }
    }
```

Here, we just change the dependency to now rely on the UserEntity class.

If we run the application after following the steps from the exercise, we will see the same output as for *Exercise 02.03*. However, if we close the application, turn on Airplane mode on the device, and reopen the app, we will still see the previously displayed information.

In this section, we have analyzed how we can persist structured data on a device and used the Room library to do so. We have also observed the interaction between Room and other libraries such as Retrofit and flows and how we can use flows to combine data streams from Room and Retrofit in a very straightforward way. In the next section, we will look at how we can persist simple data in key-value pairs.

Understanding and using the DataStore library

In this section, we will discuss how we can persist key-value pairs of data and how we can use the DataStore library for this. In Android, we have the possibility of persisting primitives and strings in key-value pairs. In the past, this was done through the `SharedPreferences` class, which was part of the Android framework. The keys and values would ultimately be saved inside an XML file on the device. Because this deals with I/O operations, it evolved over time to give the possibility to save data asynchronously and to keep an in-memory cache for quick access to data. There were, however, some inconsistencies with this, especially when the `SharedPreferences` object was initialized. DataStore is designed to address these issues because it's integrated with coroutines and flows.

To add DataStore to a project, we will need the following dependency:

```
dependencies {

    ...

    implementation "androidx.datastore:datastore-
preferences:1.0.0"

    ...

}
```

Using DataStore will look something like this:

```
private val KEY_TEXT = stringPreferencesKey("key_text")
class AppDataStore(private val dataStore:
    DataStore<Preferences>) {

    val savedText: Flow<String> = dataStore.data
        .map { preferences ->
            preferences[KEY_TEXT].orEmpty()
        }

    suspend fun saveText(text: String) {
        dataStore.edit { preferences ->
            preferences[KEY_TEXT] = text
        }
    }
}
```

The KEY_TEXT field will represent a key that will be used to store some text. DataStore<Preferences> is responsible for obtaining and writing the data to SharedPreferences. The savedText field will monitor changes in the preferences and will emit a new value for each change in a Flow object. To write data in an asynchronous way, we will need to edit the current data store and set the value associated with the key.

To initialize the DataStore library, we will need to declare the following as a top-level declaration:

```
val Context.dataStore: DataStore<Preferences> by
preferencesDataStore(name = "my_preferences")
```

This will allow us to access the DataStore library in the rest of the application.

When we want to initialize AppDataStore, we can use the following code:

```
val appDataStore = AppDataStore(dataStore)
```

This allows us to wrap the DataStore class and avoid exposing the dependencies to other places in the application.

In this section, we have looked at how we can persist data in key-value pairs and how we can use the DataStore library to do this. In the next section, we will create an Android application that will use DataStore and integrate it with Kotlin flows and coroutines.

Exercise 02.05 – Using DataStore to persist data

Modify *Exercise 02.04* and introduce the DataStore library, which will persist the number of executed requests to get the user and display this number above the list of items.

To complete the exercise, you will need to do the following:

- Create a class named AppDataStore that will manage interaction with the DataStore library.
- Modify the MainViewModel class so that the AppDataStore dependency is injected and used to retrieve the current number of requests and increment the number of requests.
- Modify the MainActivity class to add a new Text object that will display the count of requests.

Follow these steps to complete the exercise:

1. Add the following dependency to the `app/build.gradle` file:

```
dependencies {

    …

    implementation "androidx.datastore:datastore-
        preferences:1.0.0"

    …

}
```

2. Create an `AppDataStore` class, as follows:

```
private val KEY_COUNT = intPreferencesKey("key_count")
class AppDataStore(private val dataStore:
    DataStore<Preferences>) {

    val savedCount: Flow<Int> = dataStore.data
        .map { preferences ->
            preferences[KEY_COUNT] ?: 0
        }

    suspend fun incrementCount() {
        dataStore.edit { preferences ->
            val currentValue = preferences[KEY_COUNT]
                ?: 0
            preferences[KEY_COUNT] = currentValue.
                inc()
        }
    }
}}
```

Here, `KEY_COUNT` represents the key used by the DataStore library to store the number of requests. The `saveCount` field will emit a new count value every time it changes, and `incrementCount` will be increment the current saved number by 1.

3. Now, set up the `AppDataStore` dependency, just like how we handled the Retrofit and Room dependencies. The code is illustrated in the following snippet:

```
val Context.dataStore: DataStore<Preferences> by
preferencesDataStore(name = "my_preferences")
```

```
class MyApplication : Application() {

    companion object {
        ...
        lateinit var appDataStore: AppDataStore
    }

    override fun onCreate() {
        super.onCreate()
        ...
        appDataStore = AppDataStore(dataStore)
    }
}
```

Here, we initialize the DataStore object and then inject it into the AppDataStore class.

4. Next, modify the MainViewModel class, as follows:

```
class MainViewModel(
    private val userService: UserService,
    private val userDao: UserDao,
    private val appDataStore: AppDataStore
) : ViewModel() {
    var resultState by mutableStateOf(UiState())
        private set
    init {
        viewModelScope.launch {
            flow { emit(userService.getUsers()) }
                .onEach {
                    val userEntities =
                        it.map { user -> UserEntity
                            (user.id, user.name, user.
                            username, user.email) }
                    userDao.insertUsers(userEntities)
                    appDataStore.incrementCount()
                }.flatMapConcat { userDao.getUsers() }
                .catch { emitAll(userDao.getUsers()) }
```

```
                    .flatMapConcat { users ->
                        appDataStore.savedCount.map {
                            count -> UiState(users,
                                count.toString()) }
                    }
                    .flowOn(Dispatchers.IO)
                    .collect {
                        resultState = it
                    }
                }
            }
        }
```

Here, we add a new dependency to `AppDataStore`, then we call
`incrementCount` from `AppDataStore` after the users from Retrofit are
inserted, and then we will insert `savedCount` from `AppDataStore` into the
existing flow and create a new `UiState` object that contains a list of users and the
count, which will be collected in the `resultState` object.

5. The `UiState` class will look something like this:

```
data class UiState(
    val userList: List<UserEntity> = listOf(),
    val count: String = ""
)
```

This class will hold information from both of our persistent data sources.

6. Next, change `MainViewModelFactory`, as follows:

```
class MainViewModelFactory : ViewModelProvider.Factory {
    override fun <T : ViewModel> create(modelClass:
        Class<T>): T =
        MainViewModel(
            MyApplication.userService,
            MyApplication.userDao,
            MyApplication.appDataStore
        ) as T
}
```

Here, we will inject a new dependency to `AppDataStore` into the
`MainViewModel` class.

7. Finally, modify the `MainActivity` class, as follows:

```
@Composable
fun UserList(uiState: UiState) {
    LazyColumn(modifier = Modifier.padding(16.dp)) {
        item(uiState.count) {
            Column(modifier = Modifier.padding(16.dp)) {
                Text(text = uiState.count)
            }
        }
        items(uiState.userList) {
            Column(modifier = Modifier.padding(16.dp)) {
                Text(text = it.name)
                Text(text = it.username)
                Text(text = it.email)
            }
        }
    }
}
```

Here, we replaced the list of `UserEntity` with the `UiState` dependency and
added a new row in a list of items that will indicate the count of requests.

If we run the application, we will see at the top the current count of requests made to
the server. If we kill and reopen the application, then we will see that count increase,
which shows how it will survive the application being stopped by the user or killed by the
operating system.

In this section, we have analyzed another common way of persisting data on an Android
device through the DataStore library. We also observed how easy it is for the DataStore
library to be integrated with flows and other libraries such as Room and Retrofit.

Summary

In this chapter, we have looked at how we can load and persist data in Android and the rules we must follow for threading. We first analyzed how we can load data asynchronously and focused on coroutines and flows, for which we have done simple exercises for performing asynchronous operations on different threads and updating the UI on the main thread. We then studied how to load data from the internet using OkHttp and Retrofit, and followed this up with how to persist data using Room and DataStore and how we can integrate all of these with coroutines and flows. We highlighted the usage of these libraries in exercises, and we also showed how they can be integrated with coroutines and flows. The integration of different flows of data was combined in the `ViewModel` class, in which we loaded the network data and inserted it into the local database. This is generally not a good approach, and we will expand on how we can improve this in future chapters.

In the next chapter, we will look at how we can present data to the user and the libraries and frameworks we can use to achieve this.

3

Understanding Data Presentation on Android

In this chapter, we will study the libraries available for presenting data on the **user interface (UI)**. We will start this chapter by analyzing the lifecycles of activities and fragments (what responsibilities they had in the past and what responsibilities they have now) with the introduction of the `ViewModel` and `Lifecycle` libraries. We will then move on to analyze aspects of how the UI works and look at how the Jetpack Compose library revolutionized building UIs through its declarative approach. Finally, we will look at how we can navigate between different screens that are built in Compose by using the `Navigation` library with the `Compose` extension.

In this chapter, we will cover the following main topics:

- Analyzing lifecycle-aware components
- Using Jetpack Compose to build UIs

By the end of the chapter, you will become familiar with how to present data on the UI using ViewModel and Compose.

Technical requirements

Hardware and software requirements are as follows:

- Android Studio Arctic Fox 2020.3.1 Patch 3

The code files for this chapter can be found here: `https://github.com/PacktPublishing/Clean-Android-Architecture/tree/main/Chapter3`.

Check out the following video to see the Code in Action: `https://bit.ly/3lmMIOg`

Analyzing lifecycle-aware components

In this section, we will analyze the lifecycles of activities and fragments and the potential issues that are caused when working with them. We will also observe how the introduction of ViewModel and LiveData solves these problems.

When the Android operating system and its development framework were released, activities were the most commonly used components when developing an application, as they represent the entry point of the interaction between an application and a user. As technology in displays and resolutions improved, apps could then present more information and controls that the user could interact with. For developers, this meant that the code required to manage the logic for a single activity increased, especially when dealing with different layouts for landscape and portrait. The introduction of fragments was meant to solve some of these problems. Responsibilities for handling the logic in different parts of the screen could now be divided into different fragments.

The introduction of fragments, however, didn't solve all of the issues developers were dealing with, mainly because both activities and fragments have their own lifecycles. Dealing with lifecycles created the possibility of apps having context leaks, and the combination of lifecycles and inheritance made both activities and fragments hard to unit test.

The lifecycle of an activity is as follows:

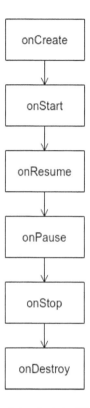

Figure 3.1 – Activity lifecycle

In *Figure 3.1*, we can see the six most well-known states of an activity:

- *CREATED*: The activity enters this state when the onCreate method is called. This will be called when the system creates the activity.

- *STARTED*: The activity enters this state when the onStart method is called. This will be called when the activity is visible to the user.

- *RESUMED*: The activity enters this state when the onResume method is called. This will be called when the activity is in focus (the user can interact with it).

The next three states are called when the activity is no longer in focus. This can be caused either by the user closing the activity, putting it in the background, or another component gaining focus:

- *PAUSED*: The activity enters this state when the onPause method is called. This will be called when the activity is visible but no longer in focus.

- *STOPPED*: The activity enters this state when the onStop method is called. This will be called when the activity is no longer visible.

- *DESTROYED*: The activity enters this state when the onDestroy method is called. This will be called when the activity is destroyed by the operating system.

When we use activities in our code, dealing with the lifecycle will look something like this:

```kotlin
class MyActivity : Activity() {

    override fun onCreate(savedInstanceState: Bundle?) {
        super.onCreate(savedInstanceState)
    }

    override fun onStart() {
        super.onStart()
    }

    override fun onResume() {
        super.onResume()
    }

    override fun onPause() {
        super.onPause()
    }

    override fun onStop() {
        super.onStop()
    }
```

```
    override fun onDestroy() {
        super.onDestroy()
    }
}
```

We can see here that we need to extend the `Activity` class and, if we want to execute a particular operation in a particular state, we can override the method associated with the state and invoke the `super` call. This represents the main reason why activities are hard to unit test. The `super` calls would cause our test not only to invoke our code but also the parent class's code. Another reason activities are hard to test is because the system is the one instantiating the class, which means that we cannot use the constructor of the class for injection and must rely on setters to inject mock objects.

An important distinction should be made between the *DESTROYED* state and garbage collection. A *DESTROYED* activity doesn't mean it will be garbage collected. A simple definition of what garbage collection means is that garbage collection is the process of deallocating memory that is no longer used. Each created object takes a certain amount of memory. When the garbage collector wants to free memory, it will look at objects that are no longer referenced by other objects. If we want to make sure that objects will be garbage collected, we will need to make sure that other objects that live longer than them will have no reference to the objects we want to be collected. In Android, we want **context** (such as activity and service) objects, or other objects with lifecycles, to be collected when their `onDestroy` methods are called. This is because they tend to occupy a lot of memory and we will end up with crashes or bugs if we end up invoking methods after `onDestroy` is called. Leaks that prevent context objects from being collected are called **context leaks**. Let's look at a simple example of this:

```
interface MyListener {

    fun onChange(newText: String)
}

object MyManager {

    private val listeners = mutableListOf<MyListener>()
```

```kotlin
    fun addListener(listener: MyListener) {
        listeners.add(listener)
    }


    fun performLogic() {
        listeners.forEach {
            it.onChange("newText")
        }
    }
}
```

Here, we have a `MyManager` class in which we collect a list of `MyListener` that will be invoked when `performLogic` is called. Note that the `MyManager` class is defined using the `object` keyword. This will make the `MyManager` class static, which means the instance of the class will live as long as the application process lives. If we want an activity to listen to when the `performLogic` method is called, we will have something like the following:

```kotlin
class MyActivity : Activity(), MyListener {

    override fun onCreate(savedInstanceState: Bundle?) {
        super.onCreate(savedInstanceState)
        MyManager.addListener(this)
    }

    override fun onChange(newText: String) {
        myTextView.setText(newText)
    }
}
```

Here, `MyListener` is implemented in `MyActivity`, and when `onChange` is called, `myTextView` will be updated. The context leak occurs here when the activity is destroyed. As `MyActivity` is a `MyListener` and a reference to it is kept in `MyManager`, which lives longer, the garbage collector will not remove the `MyActivity` instance from memory. If `performLogic` is called after `MyActivity` is destroyed, we will get `NullPointerException`, because `myTextView` will be set to null; or, if multiple instances of `MyActivity` leak, it could potentially lead to consuming the entire application's memory. A simple fix for this is to remove the reference to `MyActivity` when it is destroyed:

```kotlin
object MyManager {
    ...
    fun removeListener(listener: MyListener){
        listeners.remove(listener)
    }
    ...
}
class MyActivity : Activity(), MyListener {
    ...
    override fun onDestroy() {
        MyManager.removeListener(this)
        super.onDestroy()
    }
    ...
}
```

Here, we add a simple method to remove `MyListener` from the list and invoke it from the `onDestroy` method.

Working with fragments will lead to the same type of problems as activities. Fragments have their own lifecycle and inherit from a parent `Fragment` class, which makes them vulnerable to context leaks and hard to unit test.

The lifecycle of a fragment is as follows:

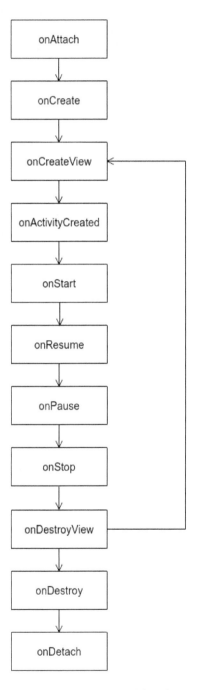

Figure 3.2 – Fragment lifecycle

In *Figure 3.2*, we can see that the fragment has similar lifecycle states to the activity. The `onAttach` and `onDetach` callbacks deal with when the fragment is attached to and detached from the activity. `onActivityCreated` is called when the activity completes its own `onCreate` call. The `onCreateView` and `onDestroyView` callbacks deal with inflating and destroying a fragment's views. One of the reasons these callbacks exist is because of the fragment back stack. This is a stack structure in which fragments are kept so that when the users press the *Back* button, the current fragment is popped out of the stack and the previous fragment is displayed. When fragments are replaced in the back stack, they aren't fully destroyed; just their views are destroyed to save memory. When they are popped back to be viewed by the user, they will not be re-created, and `onCreateView` will be called.

In order to solve the problems caused by dealing with activity and fragment lifecycles, a set of libraries was created that are part of the `androidx.lifecycle` group. The `Lifecycle` class was introduced, which is responsible for keeping the current lifecycle state and handling transitions between lifecycle events. The events and states of the `Lifecycle` class would be as follows:

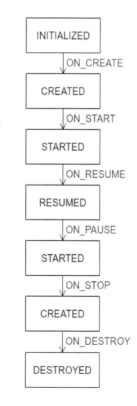

Figure 3.3 – Lifecycle states

In *Figure 3.3*, we can see that the `Lifecycle` class only has four states (*INITIALIZED*, *CREATED*, *STARTED*, and *DESTROYED*), and it will deal with six events (`ON_CREATE`, `ON_START`, `ON_RESUME`, `ON_PAUSE`, `ON_STOP`, and `ON_DESTROY`). If we wish for a certain class to be lifecycle-aware, it will need to implement the `LifecycleOwner` interface. Activities and fragments already implement this interface. We can see that for activities, the events match the existing callbacks, but for fragments, some changes are required to match these new events. The `onAttach`, `onDetach`, and `onActivityCreated` methods are deprecated, so they shouldn't be used with regard to the new `Lifecycle` library. The other change made for fragments is the introduction of a `viewLifecycleObserver` instance variable, which is used to handle the lifecycle between `onCreateView` and `onDestroyView`. This observer should be used when registering for lifecycle-aware components and you wish to update the UI.

In Android, when a configuration change (device rotation and language change, for example) occurs, then activities and fragments are re-created (the current instance is destroyed and a new instance will be created). This typically causes problems when these configuration changes occur while data is loaded or when we want to restore the previously loaded data. The `ViewModel` class is meant to solve this problem, along with the issue of testability of activities and fragments. A ViewModel will live until the activity or fragment it is connected to is destroyed and not re-created. The ViewModel comes with an `onCleared` method, which can be overwritten to clear any subscriptions to any pending operations.

ViewModels are often paired with a class called `LiveData`. This is a lifecycle-aware component that observes and emits data. The combination of the two classes eliminates the risks of context leaks, as `LiveData` will only emit data when the observer is in a *STARTED* or *RESUMED* state. An additional benefit is that it will keep the last data held; so, in the case of a configuration change, the last data kept in `LiveData` will be re-emitted. This benefit allows activities and fragments to observe the changes and restore the UI to the way it was before they were re-created. In Jetpack Compose, `LiveData` isn't necessary due to Compose's own set of state handling classes.

To use `ViewModel` and `LiveData`, you will need the following libraries to be added to `build.gradle`:

```
implementation "androidx.lifecycle:lifecycle-viewmodel-
ktx:2.4.0"
```

```
implementation "androidx.lifecycle:lifecycle-livedata-
ktx:2.4.0"
```

For integration with Jetpack Compose we will need the following:

```
implementation "androidx.lifecycle:lifecycle-viewmodel-
compose:2.4.0"
implementation "androidx.compose.runtime:runtime-livedata:2.4.0
"
```

An example of a `ViewModel` and `LiveData` implementation will look something like this:

```
class MyViewModel : ViewModel() {

    private val _myLiveData = MutableLiveData("")
    val myLiveData: LiveData<String> = _myLiveData

    init {
        _myLiveData.value = "My new value"
    }
}
```

In the preceding example, we extend the `ViewModel` class and define two `LiveData` instance variables. The `_myLiveData` variable is defined as `MutableLiveData` and is set to private. This is to prevent other objects from changing the values of `LiveData`. The `myLiveData` variable is public and can be used by `Lifecycle` owners to observe changes on `LiveData`.

To obtain the instance of a ViewModel in an activity or fragment, we can use the following:

```
class MainActivity : AppCompatActivity() {
    override fun onCreate(savedInstanceState: Bundle?) {
        super.onCreate(savedInstanceState)
        ...
        val myViewModel : MyViewModel by viewModels()
        ...
    }
}
```

Here, the `viewModels` method will retrieve the instance of `MyViewModel`. The method provides the ability to pass along a `ViewModelProvider.Factory` object. This is useful in situations where we want to inject various objects in our ViewModel. This will look something like this:

```
val myViewModel : MyViewModel by viewModels {
    object : ViewModelProvider.Factory {
        override fun <T : ViewModel>
            create(modelClass: Class<T>): T {
            return MyViewModel() as T
        }
    }
}
```

If we want to observe the changes on `LiveData`, we would need to do something like this:

```
class MainActivity : ComponentActivity() {
    override fun onCreate(savedInstanceState: Bundle?) {
        ...
        super.onCreate(savedInstanceState)
        val myViewModel: MyViewModel by viewModels()
        myViewModel.myLiveData.observe(this) { text ->
            myTextView.text = text
        }
        ...
    }
}
```

In the preceding example, we invoke the `observe` method, where we pass the activity as `LifecycleOwner` and pass a Lambda as `Observer`, which will be invoked when `LiveData` changes its value.

If we want to use `ViewModel` with `LiveData` in Jetpack Compose, we must do the following:

```
@Composable
fun MyScreen(viewModel: MyViewModel = viewModel()) {
    viewModel.myLiveData.observeAsState().value?.let {
        MyComposable(it)
```

```
        }
    }

@Composable
fun MyComposable(text: String) {
    ...
}
```

Here, we are using the `viewModel` method to obtain the `MyViewModel` instance. This method also offers the possibility of passing a `ViewModelProvider.Factory` instance, such as the previous `viewModel` method. The `observeAsState` extension method will observe changes on `LiveData` and convert them into a Compose `State` object.

In this section, we have discussed how lifecycles work in activities and fragments and the problems developers have when dealing with them. We have analyzed how the lifecycle-aware components (such as ViewModel and LiveData) solved these problems. The `ViewModel` class itself represents an implementation of the **Model-View ViewModel (MVVM)** pattern, which will be discussed in a future chapter. In the next section, we will look at an exercise in which we will use both ViewModel and LiveData and combine them with Kotlin flows.

Exercise 3.1 – Using ViewModel and LiveData

Modify *Exercise 2.5* from *Chapter 2*, *Deep Diving into Data Sources*, so that the state of the UI is kept in a `LiveData` object inside `MainViewModel`, instead of using the Compose `State` object, and display `"Total request count: x"`, where x is the number of requests at the top of the list.

To complete the exercise, you will need to build the following:

- Add the specified text in `strings.xml`.

- Create a `MainTextFormatter` class that will have one method that will return the `"Total request count: x"` text.

- Add a dependency to `MainTextFormatter` in `MainViewModel`, and pass the formatted text as a value for the `UiState.count` object.

- Remove `resultState` and replace it with a `LiveData` object.

- Update the `@Composable` functions to use `LiveData`.

Follow these steps to complete the exercise:

1. Add the `LiveData` extension library for Jetpack Compose to `app/build.gradle`:

```
implementation "androidx.compose.runtime:runtime-
livedata:$compose_version"
```

2. Add the `"Total request count"` text in `strings.xml`:

```
<string name="total_request_count">Total request
count: %d</string>
```

3. Create the `MainTextFormatter` class as follows:

```
class MainTextFormatter(private val
    applicationContext: Context) {

    fun getCounterText(count: Int) =
        applicationContext.getString(R.string.total_
request_co
    unt, count)
}
```

The reason we created this class is to prevent possible context leaks by having a `Context` object inside the `MainViewModel` class. Here, we have a method that will take a count as a parameter and return the required text.

4. Inject `MainTextFormatter` in `MainViewModel` and use the formatted text as a value for the `UiState.count` object:

```
class MainViewModel(
    ...
    private val mainTextFormatter: MainTextFormatter
) : ViewModel() {

    ...
    init {
        viewModelScope.launch {
            ...
                .flatMapConcat { users ->
                    appDataStore.savedCount.map {
                        count ->
```

```
                            UiState(
                                users,
                                mainTextFormatter.
    getCounterText(count)
                                )
                            }
                        }
                    ...
                }
            }
        }
```

5. Next, create the instance of the `MainTextFormatter` class in the `MyApplication` class:

```
class MyApplication : Application() {

    companion object {
        ...
        lateinit var mainTextFormatter:
            MainTextFormatter
    }

    override fun onCreate() {
        super.onCreate()
        ...
        mainTextFormatter = MainTextFormatter(this)
    }
}
```

6. Now, update `MainViewModelFactory` to use `MainTextFormatter`, which was just created, and pass it into `MainViewModel`:

```
class MainViewModelFactory : ViewModelProvider.Factory {
    override fun <T : ViewModel> create(modelClass:
        Class<T>): T =
        MainViewModel(
            MyApplication.userService,
            MyApplication.userDao,
```

```
                    MyApplication.appDataStore,
                    MyApplication.mainTextFormatter
            ) as T
    }
```

7. Next, add `LiveData` to `MainViewModel`:

```
class MainViewModel(
    ...
) : ViewModel() {

    private val _uiStateLiveData =
        MutableLiveData(UiState())
    val uiStateLiveData: LiveData<UiState> =
        _uiStateLiveData

    init {
        viewModelScope.launch {
            ...
                .collect {
                    _uiStateLiveData.value = it
                }
        }
    }
}
```

Here, we have defined the two `LiveData` variables, one to update the value and the other to be observed, and in the `collect` method, we update the value of `LiveData`.

8. In `MainActivity`, update the `@Composable` functions to use `LiveData`:

```
...
@Composable
fun Screen(viewModel: MainViewModel = viewModel(factory =
MainViewModelFactory())) {
    viewModel.uiStateLiveData.observeAsState().value?.let
    {
```

```
        UserList(uiState = it)
    }
}
...
```

Here, we call the `observeAsState` extension method on `LiveData` from `MainViewModel`, and then call the `UserList` method, which will redraw the UI for each new value.

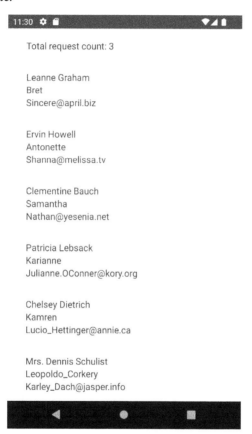

Figure 3.4 – Output of Exercise 3.1

If we run the application, we will see the same list of users, and at the top, we will see `"Total request count: x"` instead of just the x character that was there before, as shown in *Figure 3.4*. In this exercise, we used Jetpack Compose for rendering the UI. In the section that follows, we will analyze how Android handles UIs and go more in-depth into the Jetpack Compose framework.

Using Jetpack Compose to build UIs

In this section, we will analyze how to build UIs for Android applications using the `View` hierarchy and look at the implications this has for applications. We will then look at how Jetpack Compose simplifies and changes how UIs are built and how we can use Compose to create UIs. We will be looking at Jetpack Compose with the view of how we can integrate it with other libraries and how to build a simple UI. For more information on how to build more complex UIs, you can refer to the official documentation found here: `https://developer.android.com/jetpack/compose`.

The way Android deals with UIs is through the `View` hierarchy. The subclasses of `View` deal with specific UI components that the user can interact with. The hierarchy looks similar to the following diagram:

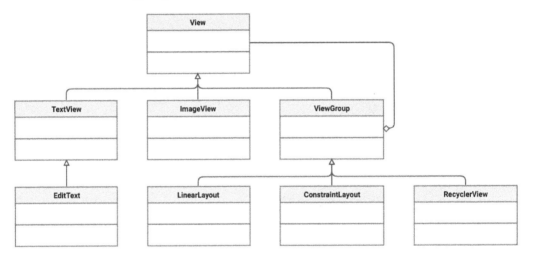

Figure 3.5 – View hierarchy

The `TextView` class deals with displaying text on the screen, `EditText` deals with handling text inputted by the user, and `Button` deals with rendering buttons on the screen. A specialized subclass of `View` is the `ViewGroup` class. This represents the base class for various layout classes that are responsible for how the views are grouped and arranged on the screen. Here, we find classes such as `LinearLayout` (which groups views one after the other either vertically or horizontally), `RelativeLayout` (which groups the views relative to the parent or to each other), or more recently, `ConstraintLayout`, which offers various ways to position views however we desire without creating many nested layouts (because it was bad for performance), which is why it became commonly used. When it comes to dealing with displaying lists of items of unknown lengths, objects such as `ListView` and `RecyclerView` are used. Both require creating adapters that will be responsible for pairing an object from a list with an associated `View` to render a row in the list in the UI.

Using `ListViews` is prone to inefficiencies caused when scrolling where views are recreated for each new row, so in a long list of items, a lot of views would be created and then garbage collected. To solve this, developers had to implement a pattern called a **ViewHolder**, which is responsible for keeping references to the views created for each row and re-using them for new rows when the user scrolls away. `RecyclerView` addresses this issue so the adapter `RecyclerView` uses requires `ViewHolder`. This means that if a user views a list of 100 items and 10 are visible on the screen, for the 10 that are visible on the screen there would be 10 views to represent each row. When the user scrolls down, the 10 views that were created at the beginning would then display the items for the currently visible items. Developers can also create custom views by extending any of the existing `View` classes. This is useful when certain UI components have to be re-used in different activities, fragments, or other custom views.

To display these views to the user, we would need to use activities and fragments. For activities, this would require invoking the `setContentView` method in the `onCreate` method, and in fragments, we would need to return a `View` object in the `onCreateView` method. We have the possibility of creating the entire layout for an activity or fragment in Java or Kotlin, but this would lead to a lot of code being written. This, and the fact that we can have different layouts for different screen sizes or device rotation, led to using the `res/layout` folder, in which we can specify how a layout might look. An example of how this might look is as follows:

```xml
<?xml version="1.0" encoding="utf-8"?>
<androidx.constraintlayout.widget.ConstraintLayout
    xmlns:android="http://schemas.android.com/apk/res/android"
    xmlns:app="http://schemas.android.com/apk/res-auto"
    android:layout_width="match_parent"
    android:layout_height="match_parent">
    <TextView
        android:id="@+id/text_view"
        android:layout_width="wrap_content"
        android:layout_height="wrap_content"
        android:text="Hello World"
        app:layout_constraintStart_toStartOf="parent"
        app:layout_constraintTop_toTopOf="parent" />
</androidx.constraintlayout.widget.ConstraintLayout>
```

In the preceding example, we define `ConstraintLayout`, which contains only `TextView` that displays a "Hello World" text. To obtain a reference to `TextView` to allow us to change the text because of an action or data being loaded, we would need to use the `findViewById` method from either the `Activity` class or the `View` class. This would look something like the following:

```
class MainActivity : ComponentActivity() {
    override fun onCreate(savedInstanceState: Bundle?) {
        super.onCreate(savedInstanceState)
        setContentView(R.layout.activity_main)
        val textView =
            findViewById<TextView>(R.id.text_view)
        textView.text = "Hello new world"
    }
}
```

This approach would lead to possible crashes within an application. Developers would need to make sure that when a layout was set for `Activity` or `Fragment` and `findViewById` was used, then the view was added to the `layout` file. With the introduction of Kotlin, this was initially addressed through the Kotlin Synthetics framework, which generated extensions for the declared views in a layout. Kotlin Synthetics would generate an extension for a View's `android:id` XML tag, which would be accessible in the code. Later, this was replaced with `ViewBinding`. When `ViewBinding` is used in a project, a class is generated for each layout that will hold references to all the views in the layout, eliminating potential crashes related to `findViewById`. All these approaches with regard to creating your UI are defined as **imperative** because we need to specify the views that our interface uses and control how we update the views when data is changed.

An alternative approach to this is the **declarative** way of creating the UI. This concept allows us to describe what we want to show on the UI and the framework by using the appropriate views based on the description we provide. The notions of **state** are introduced here, where the UIs are redrawn when states change, rather than updating the existing views. In Android, we can use Jetpack Compose to create UIs in a declarative way. We no longer have to deal with the `View` hierarchy and instead use `@Composable` functions, in which we specify what we want to display on the screen without thinking of how we need to display it, and we can also create the UI using Kotlin using less code than we would normally. In Compose, the `Hello World` example would look something like the following code:

```
class MainActivity : ComponentActivity() {
    override fun onCreate(savedInstanceState: Bundle?) {
```

```
        super.onCreate(savedInstanceState)
        setContent {
            Surface {
                HelloWorld()
            }
        }
    }
}


@Composable
fun HelloWorld() {
    Text(text = "Hello World")
}
```

If we want to update the text because of a change in data, we will need to use State objects from the Compose library. Compose will observe these states and, when the values are changed, Compose will redraw the UI associated with that state. An example of this is as follows:

```
@Composable
fun HelloWorld() {
    val text = remember { mutableStateOf("Hello World") }
    ShowText(text = text.value) {
        text.value = text.value + "0"
    }
}


@Composable
fun ShowText(text: String, onClick: () -> Unit) {
    ClickableText(
        text = AnnotatedString(text = text),
        onClick = {
            onClick()
        })
}
```

In this example, when the text is clicked, the 0 character is appended to the text and the UI is updated. This is because of the use of `mutableStateOf`. The `remember` method is needed because this state is kept inside a `@Composable` function, and it is used to keep the state intact while recomposition happens (the UI is redrawn). To make the text clickable, we needed to change from `Text` to `ClickableText`. The reason we are using two `@Composable` functions is that we want to keep the `@Composable` functions as re-usable as possible. This is called **state hoisting**, where we separate the stateful (`HelloWorld`) components from the stateless components (`ShowText`).

When it comes to rendering lists of items, Compose offers a simple way of rendering them in the form of `Column` (for when the length of the list is known and short), and `LazyColumn` (when the list of items is unknown and could potentially be long). An example of this is from *Exercise 3.1*:

```
LazyColumn(modifier = Modifier.padding(16.dp)) {
        item(uiState.count) {
            Column(modifier = Modifier.padding(16.dp)) {
                Text(text = uiState.count)
            }
        }
        items(uiState.userList) {
            Column(modifier = Modifier.padding(16.dp)) {
                Text(text = it.name)
                Text(text = it.username)
                Text(text = it.email)
            }
        }
    }
```

Here, we display a header at the top of the item list, and we use another column to set the padding for the row; then, we display the entire list of items with the use of the `items` function, and for each row, we set the padding and display a group with three texts.

If we want to display input fields and buttons, then we can look at how we implemented the UI in *Exercise 2.1*, from *Chapter 2*, *Deep Diving into Data Sources*:

```
@Composable
fun Calculator(
    a: String, onAChanged: (String) -> Unit,
    b: String, onBChanged: (String) -> Unit,
    result: String,
```

```
        onButtonClick: () -> Unit
) {
    Column(modifier = Modifier.padding(16.dp)) {
        OutlinedTextField(
            value = a,
            onValueChange = onAChanged,
            keyboardOptions = KeyboardOptions(keyboardType
                = KeyboardType.Number),
            label = { Text("a") }
        )
        OutlinedTextField(
            value = b,
            onValueChange = onBChanged,
            keyboardOptions = KeyboardOptions(keyboardType
                = KeyboardType.Number),
            label = { Text("b") }
        )
        Text(text = result)
        Button(onClick = onButtonClick) {
            Text(text = "Calculate")
        }
    }
}
```

Here, we used `OutlinedTextField` to render the equivalent of `TextInputLayout`. We could have used `TextField` if we wanted the equivalent of a simple `EditText`. For displaying a button, we can use the `Button` method, which uses `Text` for rendering the text on the button.

Compose also has integrations with other libraries, such as `ViewModel` and `LiveData`:

```
@Composable
fun Screen(viewModel: MainViewModel = viewModel(factory =
MainViewModelFactory())) {
    viewModel.uiStateLiveData.observeAsState().value?.let {
        UserList(uiState = it)
    }
}
```

Here, we can pass `ViewModel` as a parameter in our `Composable` function and use the `observeAsState` function to convert `LiveData` into a `State` object, which will then be observed by Compose to redraw the UI. Compose also supports integration with the `Hilt` library. When Hilt is added to a project, then there is no need to specify `Factory` for the ViewModel.

Another important feature of Compose is how it deals with navigation between different screens. The Compose navigation is built upon the `androidx.navigation` library. This allows Compose to use the `NavHost` and `NavController` components to navigate between different screens. The screens are built using Compose, which means that an application using only Compose would ideally have only one activity. This eliminates any potential problems regarding activity and fragment lifecycles. To introduce navigation into a project, the following library is required:

```
dependencies {

    ...

    implementation "androidx.navigation:navigation-
compose:2.4.0-rc01"

    ...

}
```

If we want to navigate from one screen to another, we will need to obtain `NavHostController` and pass it into a `@Composable` method that will represent the structure of the application:

```
Surface {
    val navController = rememberNavController()
    AppNavigation(navController = navController)
}
```

The `AppNavigation` `@Composable` method will look something like this:

```
@Composable
fun AppNavigation(navController: NavHostController) {
    NavHost(navController, startDestination = "screen1") {
        composable(route = "screen1") {
            Screen1(navController)
        }
        composable(
            route = "screen2/{param}",
```

```
                arguments = listOf(navArgument("param") { type
                    = NavType.StringType })
            ) {
                Screen2(navController,
                    it.arguments?.getString("param").orEmpty())
            }
        }
    }
```

In `AppNavigation`, we invoke the `NavHost` `@Composable` function in which we will place the screens of the application along with a route to each of them. In this case, `Screen1` will have a simple route to navigate to and `Screen2` will require an argument when it is navigated to indicated through the `{param}` notation. For arguments, we will need to specify the type of the argument. In this case, it will be `String`, and `NavType.StringType` indicates this. If we wish to pass more complex arguments, then we will need to supply our own custom types and indicate how they should be serialized and deserialized. When we want to navigate from `Screen1` to `Screen2`, then we will need to do the following:

```
@Composable
fun Screen1(navController: NavController) {
    Column(modifier = Modifier.clickable {
        navController.navigate("screen2/test")
    }) {
        Text(text = "My text")
    }
}
```

When `Column` is clicked in `Screen1`, it will invoke `NavController` to navigate to `Screen2` and pass the `test` argument. `Screen2` will look like the following:

```
@Composable
fun Screen2(navController: NavController, text: String) {
    Column {
        Text(text = text)
    }
}
```

`Screen2` will use the text extracted from `it.arguments?.getString("param").orEmpty()` and it will display it on the UI.

In this section, we have discussed how Android deals with UIs. We have looked over the imperative approach and then introduced the declarative approach for Uis. We have analyzed the Jetpack Compose library and the problems it attempts to solve, such as less code and no XML declarations for layouts. It follows the principles of libraries from other technologies (such as React and SwiftUI) and shows how UIs can be built from a functional programming point of view. In the next section, we will look at an exercise for how we can use Compose to navigate between two screens in an application.

Exercise 3.2 – Navigating using Jetpack Compose

Modify *Exercise 3.1* so that the current @Composable functions are moved into a new file named UserListScreen, then create a new file with new @Composable functions that will render a simple text called UserScreen. When a user from the list is clicked, the new screen is opened and it will display the name of the user.

To complete the exercise, you will need to build the following:

1. Create an AppNavigation sealed class that will have two variables. The first variable, named route, will be String and the second variable, named argumentName, will be String and default to empty. Two subclasses of AppNavigation will be Users (which will set the route variable to "users") and User (which will set the route to "users/{name}", then argumentName to name, and a method to create the route for a specific name).

2. In MainActivity, rename the screen @Composable function to Users, and using the NavController object, set up a click listener on the list row and navigate to the route from the User class in AppNavigation.

3. Create a new @Composable function named User, which will be responsible for showing a simple Text and will have the text displayed as a parameter.

4. In MainActivity, create a @Composable function named MainApplication, which will use the NavHost @Composable function to link the navigation between the two screens.

Follow these steps to complete the exercise:

1. Add the navigation library for Compose in app/build.gradle:

```
dependencies {

    ...

    implementation "androidx.navigation:navigation-
compose:2.4.0-rc01"

    ...

}
```

2. Create the `AppNavigation` class, which will hold the information for the routes and arguments for each of our screens:

```
private const val ROUTE_USERS = "users"
private const val ROUTE_USER = "users/%s"
private const val ARG_USER_NAME = "name"

sealed class AppNavigation(val route: String, val
    argumentName: String = "") {

    object Users : AppNavigation(ROUTE_USERS)

    object User : AppNavigation
        (String.format(ROUTE_USER, "{$ARG_USER_NAME}")
            , ARG_USER_NAME) {

        fun routeForName(name: String) =
            String.format(ROUTE_USER, name)
    }
}
```

As the navigation relies on URLs to identify the different screens, we can take advantage of sealed classes and objects in Kotlin to keep track of the required inputs for each screen.

3. Rename the screen `@Composable` function to `Users` in `MainActivity` and add `NavController` as a parameter:

```
@Composable
fun Users(
    navController: NavController,
    viewModel: MainViewModel = viewModel(factory =
        MainViewModelFactory())
) {
    viewModel.uiStateLiveData.observeAsState().value?.let
    {
        UserList(uiState = it, navController)
    }
}
```

4. Next, pass the `NavController` parameter to `UserList` and implement the click listener for the user row:

```
@Composable
fun UserList(uiState: UiState, navController:
NavController) {
    LazyColumn(modifier = Modifier.padding(16.dp)) {
        item(uiState.count) {
            Column(modifier = Modifier.padding(16.dp)) {
                Text(text = uiState.count)
            }
        }
        items(uiState.userList) {
            Column(modifier = Modifier
                .padding(16.dp)
                .clickable {
                    navController.navigate
                        (AppNavigation.User.routeForName
                            (it.name))
                }) {
                Text(text = it.name)
                Text(text = it.username)
                Text(text = it.email)
            }
        }
    }
}
```

5. Create the `User` `@Composable` function in `MainActivity`:

```
@Composable
fun User(text: String) {
    Column {
        Text(text = text)
    }
}
```

6. Now, create an App @Composable function that will use NavHost to set up the navigation between the two screens in MainActivity:

```
@Composable
fun App(navController: NavHostController) {
    NavHost(navController, startDestination =
        AppNavigation.Users.route) {
        composable(route = AppNavigation.Users.route) {
            Users(navController)
        }
        composable(
            route = AppNavigation.User.route,
            arguments = listOf(navArgument
                (AppNavigation.User.argumentName) {
                type = NavType.StringType
            })
        ) {
            User(it.arguments?.getString(AppNavigation.
User.argumentName).orEmpty())
        }
    }
}
```

7. Finally, invoke the App function when the Activity content is set in MainActivity:

```
class MainActivity : ComponentActivity() {
    override fun onCreate(savedInstanceState: Bundle?) {
        super.onCreate(savedInstanceState)
        setContent {
            Exercise0302Theme {
                // Replace this with your application's theme
                Surface {
                    val navController =
                        rememberNavController()
                    App(navController = navController)
                }
            }
```

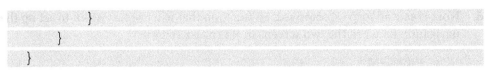

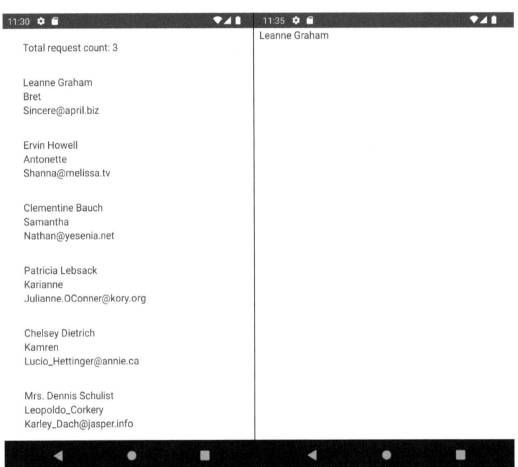

Figure 3.6 – Output of Exercise 3.2

If we run the application, we should see the same list of users as before, and if we click on a user, it will transition to a new screen that will display the selected user's name, as shown in *Figure 3.6*. If we press the *Back* button, we should see the initial list of users; that's because, by default, the `navigation` library handles back navigation.

In this exercise, we have analyzed how we can use Jetpack Compose to navigate between two screens in an application. In future chapters, we will revisit navigation when we must navigate between different screens in different modules.

Summary

In this chapter, we have analyzed how data can be presented in Android and discussed the libraries we have available now. We have looked at Android lifecycles and the potential issues that applications could have regarding lifecycles and then looked at how libraries such as `ViewModel` and `LiveData` solve most of these problems. We then looked at how the UI works in Android and how we would need to deal with using XML to define layouts in which we would insert the views that the layouts needed to display, and how we would need to update the state of the views when the data changes. We then looked at how Jetpack Compose solves these issues in a declarative functional way. We built upon the exercises in the previous chapter to show how we can integrate multiple libraries in a single application and display data from the internet.

In the next chapter, we will deal with managing the dependencies inside an application and the libraries available for doing so.

4

Managing Dependencies in Android Applications

In this chapter, we will analyze the concept of **dependency injection** (**DI**) and the benefits it provides and look at how this was done in the past in Android applications either through manual injection or using Dagger 2. We will go over some of the libraries used in Android applications, stopping and looking in more detail at the Hilt library and how it simplifies DI for an Android application.

In this chapter, we will cover the following topics:

- Introduction to DI
- Using Dagger 2 to manage dependencies
- Using Hilt to manage dependencies

By the end of this chapter, you will be familiar with the DI pattern and libraries such as Dagger and Hilt, which can be used to manage dependencies in Android applications.

Technical requirements

The hardware and software requirements are as follows:

- Android Studio Arctic Fox 2020.3.1 Patch 3

The code files for this chapter can be found here: `https://github.com/PacktPublishing/Clean-Android-Architecture/tree/main/Chapter4`.

Check out the following video to see the Code in Action: `https://bit.ly/38yFDHz`

Introduction to DI

In this section, we will look at what DI is, the benefits it provides, and how this concept is applied to an Android application. We will then look at some DI libraries and how they work.

When a class depends on functionality from another class, a dependency is created between the two classes. To invoke the functionality on the class you depend on, you will need to instantiate it, as in the following example:

```
class ClassA() {

    private val b: ClassB = ClassB()

    fun executeA() {
        b.executeB()
    }
}

class ClassB() {

    fun executeB() {

    }
}
```

In this example, `ClassA` creates a new instance of `ClassB`, and then when `executeA` is invoked, it will invoke `executeB`. This poses a problem because `ClassA` will have the extra responsibility of creating `ClassB`. Let's see what happens if `ClassB` needs to change to something such as the following:

```
class ClassB(private val myFlag: Boolean) {

    fun executeB() {
        if (myFlag) {
            // Do something
        } else {
            // Do something else
        }
    }
}
```

Here, we added the `myFlag` variable to `ClassB`, which is used in the `executeB` method. This change would cause a compile error because now `ClassA` will need to be modified to make the code compile.

```
class ClassA() {

    private val b: ClassB = ClassB(true)

    fun executeA() {
        b.executeB()
    }
}
```

Here, we will need to supply a Boolean value when we create `ClassB`.

Making these types of changes to an application as its code base increases will make it hard to maintain. A solution to this problem is to separate how we use dependencies and how we create them and delegate the creation to a different object. Continuing from the preceding example, we can rewrite `ClassA` as the following:

```
class ClassA(private val b: ClassB) {

    fun executeA() {
```

```
        b.executeB()
    }
}
```

Here, we removed the instantiation of `ClassB` and moved the variable in the constructor of `ClassA`. Now, we can create a class that will be responsible for creating the instances of both classes that looks like the following:

```
class Injector() {

    fun createA(b: ClassB) = ClassA(b)

    fun createB() = ClassB(true)

}
```

Here, we have a new class that will create an instance of `ClassA` with `ClassB` as a parameter and a separate method for creating an instance of `ClassB`. Ideally, when the program is initialized, we would need to initialize all the dependencies and pass them appropriately:

```
fun main(args : Array<String>) {
    val injector = Injector()
    val b = injector.createB()
    val a = injector.createA(b)
}
```

Here, we created `Injector`, which is responsible for creating our instances, and then invoked the appropriate methods on `Injector` to retrieve the appropriate instances of each class. What we have done here is called DI. Instead of `ClassA` creating the instance of `ClassB`, it will have an instance of `ClassB` injected through the constructor, also known as *constructor injection*.

In `ClassB`, we have an `if-else` statement in the `executeB` method. We can introduce an abstraction there, so we split the `if-else` statement into two separate implementations:

```
class ClassA(private val b: ClassB) {

    fun executeA() {
        b.executeB()
```

```
        }
}

interface ClassB {

    fun executeB()
}

class ClassB1() : ClassB {

    override fun executeB() {
        // Do something
    }
}

class ClassB2() : ClassB {

    override fun executeB() {
        // Do something else
    }
}
```

Here, ClassA remains the same and ClassB has become an interface with two implementations, called ClassB1 and ClassB2, representing the implementations of the if-else branch. Here, we can use the Injector class as well to inject one of the two implementations without requiring any change on ClassA:

```
class Injector() {

    fun createA(b: ClassB) = ClassA(b)

    fun createB() = ClassB1()

}
```

In the `createB` method, we return an instance of `ClassB1`, which will then be later injected into `ClassA`. This represents another benefit of DI, where we can make our code depend on abstractions rather than concretions and provide different concretions for different purposes. Based on this, we can define the following roles when it comes to DI:

- **Service**: Represents the object that contains useful functionality (`ClassB1` and `ClassB2` in our example)

- **Interface**: Represents the service abstraction (`ClassB` in our example)

- **Client**: Represents the object that depends on the service (`ClassA` in our example)

- **Injector**: Represents the object responsible for constructing the services and injecting them into the client (`Injector` in our example)

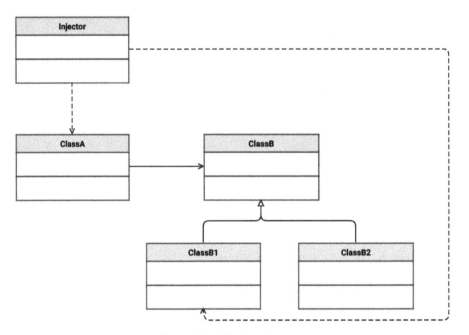

Figure 4.1 – DI class diagram

The preceding figure shows the class diagram of our example and the DI pattern. We can observe how the `Injector` class is responsible for creating and injecting the dependencies, `ClassA` is the client receiving a dependency to `ClassB`, which is the interface, and `ClassB1` and `ClassB2` represent the services.

There are multiple classifications of the types of DI, and they mainly revolve around two ways of injecting dependencies:

- **Constructor injection**: Where dependencies are passed through the constructor.

- **Field injection**: Where dependencies are passed through setter methods or by changing the instance variables. This can also be referred to as **setter injection** and it can also be expanded to **interface injection** in which the setter method is abstracted to an interface.

Another benefit of DI is the fact that it makes the code more testable. When dependencies are injected into an object, it makes the class easier to test, because in the test code, we can inject objects that allow us to mimic various behaviors, called **mocks**.

In this section, we have introduced the DI pattern, how it works, and the problems it is solving. Developers can manage an application's dependencies and injection manually, by setting up injectors. But as an application grows, it becomes hard to maintain, especially when we want certain objects to live only as long as other objects and not as long as the application, or handle different instances of the same class. There are various DI frameworks and libraries that can manage all these cases and in Android, one of the most commonly used ones is Dagger 2.

Using Dagger 2 to manage dependencies

In this section, we will analyze the Dagger 2 library, how it handles DI, how it works, how it is integrated into an Android application, and what issues it might create.

The Dagger 2 library relies on code generation based on annotation processing, which will generate the boilerplate code that is required to perform DI. The library is written in Java, and it is used for various projects outside of Android applications. Because it is written in Java, it provides compatibility for apps written in Java, Kotlin, or both. The library is built using **Java Specification Requests (JSR) 330**, which provide a set of useful annotations for DI (@Inject, @Named, @Qualifier, @Scope, and @Singleton).

When integrating Dagger 2, there are three main concepts that we will need to consider:

- **Provider**: This is represented by the classes responsible for providing the dependencies, using the @Module annotation for the classes and @Provides for the methods. To avoid many @Module definitions, we can use the @Inject annotation on a constructor, which will provide the object as a dependency.

- **Consumer**: This is represented by the classes where the dependencies are required using the @Inject annotation.

- **Connector**: This is represented by the classes that connect the providers with the consumers and is annotated with the @Component annotation.

In order to add Dagger 2 to an Android application, you will first need to add the Kotlin annotation processor plugin to the `build.gradle` file of the module in which Dagger 2 is used:

```
plugins {
    ...
    id 'kotlin-kapt'
    ...
}
```

Here, we added the `kotlin-kapt` plugin to allow Dagger 2 to generate the code necessary for DI. Next, we will need the Dagger 2 dependencies:

```
dependencies {
    ...
    implementation 'com.google.dagger:dagger:2.40.5'
    kapt 'com.google.dagger:dagger-compiler:2.40.5'
    ...
}
```

Here, we are adding a dependency to the Dagger 2 library and a dependency to the annotation processing library, which has the role of code generation. The library version should ideally be the latest stable one available in the library repository.

Let's now re-introduce the example from the previous section:

```
class ClassA(private val b: ClassB) {

    fun executeA() {
        b.executeB()
    }
}

interface ClassB {

    fun executeB()
}

class ClassB1() : ClassB {
```

```
        override fun executeB() {
            // Do something
        }
    }


class ClassB2() : ClassB {

        override fun executeB() {
            // Do something else
        }
    }
```

Here, we have the same classes with the same dependencies. Instead of defining an Injector class, we can use Dagger 2 to define an @Module:

```
@Module
class ApplicationModule {

    @Provides
    fun provideClassA(b: ClassB): ClassA = ClassA(b)

    @Provides
    fun provideClassB(): ClassB = ClassB1()
}
```

Here, we annotated the class with @Module and for each instance, we used the @Provides annotation. We can further simplify this with the @Inject annotation and delete the @Provides methods from ApplicationModule:

```
class ClassA @Inject constructor(private val b: ClassB) {
    ...
}


class ClassB1 @Inject constructor() : ClassB {
    ...
}


class ClassB2 @Inject constructor() : ClassB {
```

```
    ...
}
```

In the preceding code, we have added `@Inject` for each constructor. In the case of `ClassA`, it will have both the role of injecting `ClassB` and providing `ClassA` to other objects as a dependency. There is, however, an issue because `ClassA` has a dependency on the abstraction rather than the concretion, so Dagger will not know which instance to provide to `ClassA`. We can now add an `@Binds` annotated method to `ApplicationModule`, which will connect the abstraction with the implementation:

```
@Module
abstract class ApplicationModule {

    @Binds
    abstract fun bindClassB(b: ClassB1): ClassB

}
```

Here, we added the `bindClassB` abstract method, which is annotated with `@Binds`. This method will tell Dagger 2 to connect the `ClassB1` implementation with the `ClassB` abstraction. To avoid large `@Provides` annotations, we should try to use the annotation for dependencies where we cannot modify the code and instead rely on `@Inject` on the constructors and using `@Binds` where possible.

Now, we will need to create the connector:

```
@Singleton
@Component(modules = [ApplicationModule::class])
interface ApplicationComponent
```

Here, we are defining an `@Component` in which we specify the module the application will use. The `@Singleton` annotation tells Dagger that all the dependencies in this component will live as long as the application. At this point, we should trigger a build on the application. This will trigger the compilation, which will generate a `DaggerApplicationComponent` class. This is an implementation of `ApplicationComponent` that Dagger 2 will handle. This class will be used to create the entire dependency graph. In Android, we need an entry point for this, which is represented by the `Application` class:

```
class MyApplication : Application() {

    lateinit var component: ApplicationComponent
```

```
override fun onCreate() {
    super.onCreate()
    component = DaggerApplicationComponent.create()
    }
}
```

Here, in the MyApplication class, we are using DaggerApplicationComponent and creating the dependency graph. This will go over all the modules in the graph and invoke all the @Provides methods. The @Component annotation has another role, which is to define member injection when constructor injection is not possible. In Android, this situation occurs when dealing with life cycle components such as activities and fragments, because we are not allowed to modify the default constructors of these classes. To do this, we can do the following:

```
@Singleton
@Component(modules = [ApplicationModule::class])
interface ApplicationComponent {

    fun inject(mainActivity: MainActivity)
}
```

In ApplicationComponent, we add a method called inject and the Activity where we want the injection to be performed. In the MainActivity class, we will need to do the following:

```
class MainActivity : AppCompatActivity() {

    @Inject
    lateinit var a: ClassA

    override fun onCreate(savedInstanceState: Bundle?) {
        super.onCreate(savedInstanceState)
        setContentView(R.layout.activity_main)
        (application as
            MyApplication).component.inject(this)
        a.executeA()
    }
}
```

Here, we will need to access the `ApplicationComponent` instance created in `MyApplication` and then invoke the `inject` method from `ApplicationComponent`. This will then initialize variable a with the instance Dagger 2 created. This approach has a problem, however, because all the dependencies will live as long as the application. This means that Dagger 2 will need to keep dependencies in memory when they are not required. Dagger 2 offers a solution for this in the form of scopes and subcomponents. We can create a new Scope, which will tell Dagger 2 to only keep certain dependencies as long as an Activity is alive, and then apply this Scope to a Subcomponent, which will handle a smaller graph of dependencies.

```
@Scope
@MustBeDocumented
@kotlin.annotation.Retention(AnnotationRetention.RUNTIME)
annotation class ActivityScope
```

Here, we created a new `@Scope` annotation, which will indicate that dependencies will live as long as activities. We will next use `@ActivityScope` to create an `@Subcomponent` annotated class:

```
@ActivityScope
@Subcomponent(modules = [ApplicationModule::class])
interface MainSubcomponent {

    fun inject(mainActivity: MainActivity)

}
```

Here, we have defined a subcomponent that will use `ApplicationModule` and has an `inject` method for field injection into `MainActivity`. After that, we will need to tell Dagger 2 to create `MainSubcomponent`, by modifying `ApplicationComponent`:

```
@Singleton
@Component
interface ApplicationComponent {

    fun createMainSubcomponent(): MainSubcomponent
}
```

Here, we have removed `ApplicationModule` from `@Component` and replaced the `inject` method with a `createMainSubcomponent` method, which will allow Dagger to create `MainSubcomponent`. Finally, we will need to access `MainSubcomponent` in `MainActivity` and inject the `ClassA` dependency:

```kotlin
class MainActivity : AppCompatActivity() {

    @Inject
    lateinit var a: ClassA

    override fun onCreate(savedInstanceState: Bundle?) {
        super.onCreate(savedInstanceState)
        setContentView(R.layout.activity_main)
        (application as MyApplication).component.
            createMainSubcomponent().inject(this)
        a.executeA()
    }

}
```

Here, we access the `ApplicationComponent` instance from `MyApplication`, then create `MainSubcomponent` and then inject the `ClassA` dependency into the a variable. The code generated by Dagger 2 can be seen in the `{module}/build/generated/source/kapt/{build type}` folder and will look something similar to the following figure:

Figure 4.2 – Generated Dagger Classes

In the preceding figure, we can see Dagger will generate the implementation for the `ApplicationComponent` interface as well as the `MainSubcomponent` implementation. For dependencies that will need to be injected, it will generate a `Factory` class to create the dependency. Where we are injecting through the members, it will create an `Injector` class, which will be responsible for setting the value on the member variable, like the `MainActivity` class.

In this section, we have discussed the Dagger 2 library and how it can be used to provide and inject dependencies. Because it is a library used in other frameworks other than Android, it requires specific workarounds for injecting in activities and fragments, using member injectors and Subcomponents. An attempt at fixing this was through the introduction of the Dagger Android library, which handled the creation of `@Subcomponent` annotated classes and introduced new annotations to indicate how Subcomponents should be created. More recently, the introduction of the Hilt library was more effective at solving these problems by further simplifying the amount of code developers needed to write and providing better compatibility with components such as ViewModel. In the section that follows, we will look at the Hilt library and how it solves these problems.

Using Hilt to manage dependencies

In this section, we will discuss the Hilt DI library, how we can use it in an Android application, and the extra features it provides on top of the Dagger 2 library.

Hilt is a library built on top of Dagger 2 with a specific focus on Android applications. This is to remove the extra boilerplate code that was required to use Dagger 2 in an application. Hilt removes the need to use `@Component` and `@Subcomponent` annotated classes and in turn offers new annotations:

- When injecting dependencies in Android classes, we can use `@HiltAndroidApp` for `Application` classes, `@AndroidEntryPoint` for activities, fragments, services, broadcast receivers, and views, and `@HiltViewModel` for `ViewModels`.

- When using the `@Module` annotation, we now have the option to use `@InstallIn` and specify an `@DefineComponent` annotated class, which represents the component the module will be added to. Hilt provides a set of useful components to install modules in:

  - `@SingletonComponent`: This will make the dependencies live as long as the application.

  - `@ViewModelComponent`: This will make the dependencies live as long as a `ViewModel`.

- @ActivityComponent: This will make the dependencies live as long as an Activity.

- @FragmentComponent: This will make the dependencies live as long as a Fragment.

- @ServiceComponent: This will make the dependencies live as long as a Service.

In order to use Hilt in a project, it will require a Gradle plugin, which will need to be added as a dependency to the root build.gradle file in the project:

```
buildscript {
    repositories {
        ...
    }
    dependencies {
        ...
        classpath 'com.google.dagger:hilt-android-gradle-
            plugin:2.40.5'
    }
}
```

We will then need to add the annotation processor plugin and the Hilt plugin to the build.gradle file of the Gradle module that we want to use the Hilt library in:

```
plugins {
    ...
    id 'kotlin-kapt'
    id 'dagger.hilt.android.plugin'
}
```

The combination of these two plugins is what allows Hilt to generate the necessary source code for injecting the dependencies. Finally, we will need to add the dependency to the Hilt library:

```
dependencies {
    ...
    implementation 'com.google.dagger:hilt-android:2.40.5'
    kapt 'com.google.dagger:hilt-compiler:2.40.5'
    ...
}
```

Here, we need the dependency on the library itself and a dependency on the annotation processor like how it was necessary for Dagger 2.

Let's now re-introduce the example from the previous section:

```
class ClassA @Inject constructor(private val b: ClassB) {

    fun executeA() {
        b.executeB()
    }
}

interface ClassB {

    fun executeB()
}

class ClassB1 @Inject constructor() : ClassB {

    override fun executeB() {
        // Do something
    }
}

class ClassB2 @Inject constructor() : ClassB {

    override fun executeB() {
        // Do something else
    }
}
```

Here, we can keep the same structure of our classes and use the @Inject annotation like previously. The @Module annotated class that will provide these dependencies will look similar to a Dagger 2 Module:

```
@Module
@InstallIn(SingletonComponent::class)
abstract class ApplicationModule {
```

```
    @Binds
    abstract fun bindClassB(b: ClassB1): ClassB

}
```

In the `ApplicationModule` class, we keep the same implementation as before but now we have added the `@InstallIn` annotation, which will make the dependencies provided by this module live as long as the application will. Next, we will need to trigger the generation of components:

```
@HiltAndroidApp
class MyApplication : Application()
```

Here, we no longer need to use `DaggerApplicationComponent` to manually trigger the creation of the dependency graph and instead use `@HiltAndroidApp`, which will do this for us, as well as providing the ability to inject dependencies into the `MyApplication` class. Finally, we will need to inject the dependencies into an `Activity`:

```
@AndroidEntryPoint
class MainActivity : AppCompatActivity() {

    @Inject
    lateinit var a: ClassA

    override fun onCreate(savedInstanceState: Bundle?) {
        super.onCreate(savedInstanceState)
        setContentView(R.layout.activity_main)
        a.executeA()
    }
}
```

Here, we use the @AndroidEntry point to inform Hilt that we want to inject a dependency into an Activity and then use the @Inject annotation like how it worked in Dagger 2. The code generated by Hilt will look similar to the following figure and can be found in {module}/build/generated/source/kapt/{build type}:

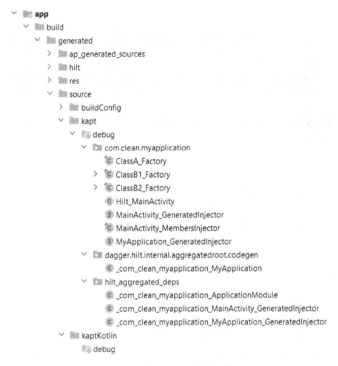

Figure 4.3 – Generated Hilt Classes

In the preceding figure, we can see Factory classes like the ones that Dagger 2 generates, but extra classes that Hilt will generate to handle the boilerplate work that was required to work with Dagger 2, such as handling the injection into activities and fragments or creating the dependency graph in the Application class.

In this section, we discussed the Hilt library, how we can use it to manage dependencies in an Android application, and how it removes boilerplate code that Dagger 2 required. In the following section, we will look at an exercise on integrating Hilt into an application together with other libraries.

Exercise 04.01 – Using Hilt to manage dependencies

Modify *Exercise 03.02 – navigating using Jetpack Compose* from *Chapter 3, Understanding Data Presentation on Android*, so that it will use Hilt to manage the dependencies across the application.

To complete the exercise, you will need to do the following:

1. Add the Hilt library to the project.

2. Create a `NetworkModule` class that will provide the Retrofit dependencies.

3. Create a `PersistenceModule` class that will provide the Room and Data Store dependencies.

4. Clean up the `MyApplication` class, delete the `MainViewModelFactory` class, and instead use the `@HiltViewModel` annotation.

5. Modify `MainActivity` to instead obtain an instance of the `MainView` model from the Hilt Compose Navigation library.

Follow these steps to complete the exercise:

1. Add the Hilt Gradle plugin to the root project `build.gradle` file:

    ```
    buildscript {
        repositories {
            ...
        }
        dependencies {
            ...
            classpath 'com.google.dagger:hilt-android-
                gradle-plugin:2.40.5'
        }
    }
    ```

2. Apply the Gradle plugin to the `build.gradle` file in the app module:

    ```
    plugins {
        ...
        id 'dagger.hilt.android.plugin'
    }
    ```

3. Add the Hilt library dependency to the app module's `build.gradle` file:

    ```
    dependencies {
        ...
        implementation 'com.google.dagger:hilt-android
            :2.40.5'
        kapt 'com.google.dagger:hilt-compiler:2.40.5'
    ```

```
implementation 'androidx.hilt:hilt-navigation-
    compose:1.0.0-rc01'
...
}
```

Here, we added a dependency that allows Hilt to work with the Jetpack Compose Navigation library.

4. Create a `NetworkModule` class in which the networking dependencies are provided:

```
@Module
@InstallIn(SingletonComponent::class)
class NetworkModule {

    @Provides
    fun provideOkHttpClient(): OkHttpClient =
        OkHttpClient
        .Builder()
        .readTimeout(15, TimeUnit.SECONDS)
        .connectTimeout(15, TimeUnit.SECONDS)
        .build()

    @Provides
    fun provideMoshi(): Moshi = Moshi.Builder().
        add(KotlinJsonAdapterFactory()).build()

    @Provides
    fun provideRetrofit(okHttpClient: OkHttpClient,
        moshi: Moshi): Retrofit = Retrofit.Builder()
        .baseUrl("https://jsonplaceholder.typicode.com/")
        .client(okHttpClient)
        .addConverterFactory(MoshiConverterFactory.create
(moshi))
        .build()

    @Provides
    fun provideUserService(retrofit: Retrofit):
```

```
        UserService =
            retrofit.create(UserService::class.java)
    }
```

Here, we have moved all the dependencies for networking and split them into separate methods for OkHttplClient, Moshi, Retrofit, and finally, the UserService class.

5. Next, create a PersistenceModule class, which will return all the persistence-related dependencies:

```
val Context.dataStore: DataStore<Preferences> by
preferencesDataStore(name = "my_preferences")
@Module
@InstallIn(SingletonComponent::class)
class PersistenceModule {

    @Provides
    fun provideAppDatabase(@ApplicationContext
        context: Context): AppDatabase =
        Room.databaseBuilder(
            context,
            AppDatabase::class.java, "my-database"
        ).build()

    @Provides
    fun provideUserDao(appDatabase: AppDatabase):
        UserDao = appDatabase.userDao()

    @Provides
    fun provideAppDataStore(@ApplicationContext
        context: Context) = AppDataStore
            (context.dataStore)
}
```

Here, we have moved all the Room-related classes and the Data Store classes. For `DataStore`, we are required to declare the `Context.dataStore` file at the top level of the file, so we will need to keep it here. The usage of `@ApplicationContext` is meant to denote that the `Context` object is represented by context of the application and not other Context objects such as an `Activity` object or `Service` object. The annotation is a `Qualifier`, which is meant to distinguish between different instances of the same class (in this case, it's to distinguish between the application context and activity context).

6. Add the `@Inject` annotation to the constructor of the `MainTextFormatter` class:

```
class MainTextFormatter @Inject constructor(@
ApplicationContext private val applicationContext:
Context) {


    fun getCounterText(count: Int) =
        applicationContext.getString(R.string.total_
            request_count, count)

}
```

This will let Hilt provide a new instance of `MainTextFormatter` every time it will be used as a dependency. Here, again, we will need to use the `@ApplicationContext` annotation to use the application `Context` object.

7. Delete all the dependencies in the `MyApplication` class and add the `@HiltAndroidApp` annotation:

```
@HiltAndroidApp
class MyApplication : Application()
```

8. Delete the `MainViewModelFactory` class.

9. Add the `@HiltViewModel` annotation to the `MainViewModel` class and `@Inject` to the constructor:

```
@HiltViewModel
class MainViewModel @Inject constructor(
    private val userService: UserService,
    private val userDao: UserDao,
    private val appDataStore: AppDataStore,
    private val mainTextFormatter: MainTextFormatter
) : ViewModel() {
```

```
    ...
}
```

10. Delete the reference to `MainViewModelFactory` in the `Users` `@Composable` method in `MainActivity`:

```
@Composable
fun Users(
    navController: NavController,
    viewModel: MainViewModel
) {
    ...
}
```

11. Change the `@Composable` App method in `MainActivity` so that it provides a `MainViewModel` instance when it invokes the `Users` method:

```
@Composable
fun App(navController: NavHostController) {
    NavHost(navController, startDestination =
AppNavigation.Users.route) {
        composable(route = AppNavigation.Users.route) {
            Users(navController, hiltViewModel())
        }
        composable(
            route = AppNavigation.User.route,
            arguments = listOf(navArgument
                (AppNavigation.User.argumentName) {
                type = NavType.StringType
            })
        ) {
            User(it.arguments?.getString(AppNavigation.
User.
    argumentName).orEmpty())
        }
    }
}
```

Here, we are using the `hiltViewModel` method, which is from the Hilt compatibility library with the Navigation library.

12. Add the `@AndroidEntryPoint` annotation to `MainActivity`:

```
@AndroidEntryPoint
class MainActivity : ComponentActivity() {

    ...

}
```

13. If you encounter a `Records requires ASM8` error when building the application, then add the following to the root project's `gradle.properties` file:

```
android.jetifier.ignorelist=moshi-1.13.0
```

This error is caused by an incompatibility that currently exists in the Android build tools and should be resolved when later updates are available.

If we run the application covered in this exercise, the functionality and user interface should remain the same as before. The role of Hilt here was to simplify how we manage dependencies, shown by how we have simplified the `MyApplication` class, leaving it with a simple annotation, and the fact that we have removed `MainViewModelFactory`, which itself had to depend on the `MyApplication` class. We can also see how easy it is to integrate Hilt with the rest of the libraries we used in the exercise.

Summary

In this chapter, we looked at the DI pattern and some of the more popular libraries that are available to apply this pattern to an Android application. We looked initially at Dagger 2 and how it can be integrated into an application, and then we analyzed the Hilt library, which is built on top of Dagger 2 and solves further problems that are specific to Android development.

There are other libraries that can be used to manage dependencies, such as Koin, which uses the Service Locator pattern (in which a registry is created and dependencies can be obtained) and is developed for Kotlin development. The exercise in this chapter showed how Hilt can be integrated with other libraries into an Android application. The problem is that the application still has no shape; there isn't anything we can point to that indicates what the use cases are. In the chapters that follow, we will look further into how we can structure our code to give it a shape using the Clean Architecture principles, starting with defining entities and use cases.

Part 2 – Domain and Data Layers

In this part, you will learn about the data and domain layers, how to structure your code into these two layers, and what components are involved in assembling them.

This part includes the following chapters:

5

Building the Domain of an Android Application

In this chapter, we will analyze what the architecture of an Android application typically looks like and its three main layers (**presentation**, **domain**, and **data**). Then, we will learn how we can translate it into clean architecture and focus on the domain layer, which sits at the center of the architecture. Next, we will look at the role it plays in the architecture of an application and what its entities and use cases are. Finally, we will look at an exercise, in which we are going to see how we can set up an Android Studio project with multiple modules and use them to structure the domain layer.

In this chapter, we will cover the following topics:

- Introducing the app's architecture
- Creating the domain layer

By the end of this chapter, you will be familiar with the domain layer of an application, domain entities, and use cases.

Technical requirements

These are the hardware and software requirements:

- Android Studio – Arctic Fox | 2020.3.1 Patch 3

The code files for this chapter can be found here: `https://github.com/PacktPublishing/Clean-Android-Architecture/tree/main/Chapter5`.

Check out the following video to see the Code in Action: `https://bit.ly/3826FH6`

Introducing the app's architecture

In this section, we will discuss the most common architecture that can be applied to an Android application and how it can be combined with **clean architecture** principles, and see how we should ideally structure our code base.

In the exercises from the previous chapters, we saw how, for an application that requires the integration of multiple data sources for networking and persistence, we had to put a lot of logic inside the `ViewModel` class. In those examples, `ViewModel` had multiple responsibilities, including fetching the data from the internet, persisting it locally, and holding the required information in the user interface. On top of these extra responsibilities, `ViewModel` also had many dependencies on the different data sources; this means that a change in the networking or persistence libraries would require a change in `ViewModel`. To solve this problem, our code would need to be split into separate layers with different responsibilities. Typically, the layers would look like the following figure:

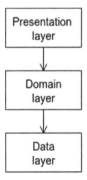

Figure 5.1 – An app architecture diagram

In *Figure 5.1*, we can see that there are three layers with different responsibilities:

- **Presentation layer**: This layer is responsible for displaying data on the screen (also known as the **UI layer**). This usually contains classes required for managing the user interface and classes that will perform logic related to the user interface, such as `ViewModels`.

- **Domain layer**: This layer is responsible for fetching data from the data layer and performing business logic that can be reused across an app.

- **Data layer**: This layer is responsible for handling the business logic of an application that deals with the managing of data.

We can apply the *clean architecture* principles on top of the layered architecture by placing the domain layer at the center, as shown in *Figure 5.2*, and making it the place to store our *entities* and *use cases*. At the outer layers are the presentation and data layers, which are represented by the **interface adapter layer** (represented by `ViewModels` and `Repositories`) and the **framework layer** (represented by the user interface and persistence and networking frameworks):

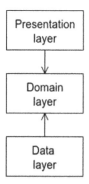

Figure 5.2 – An app layer dependency diagram

In the preceding figure, we can see that the dependencies between the domain layer and the data layer are inverted. The domain layer will still draw data from the data layer, but because it has inverted dependencies, it will be less impacted by any changes to that layer, just as if any changes occur to the presentation layer, they will not impact the domain layer. If the app suffers from any changes to the use cases, then it will drive the changes in both the presentation and data layer.

To separate the layers, we can use **Android modules**. This will help us impose further rigor on the project by avoiding unwanted dependencies between the layers. This also helps improve build times in large applications because of Gradle build caching, which will only rebuild modules that had code changes. This will look something like the following figure:

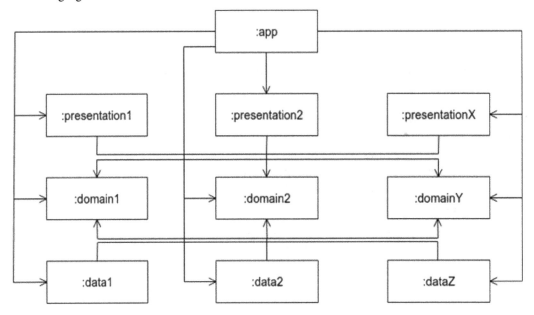

Figure 5.3 – An app module diagram

We can see that there isn't a limited number of modules we need for each layer or that we should have corresponding modules between the three layers. The expansion of each layer can be driven by different factors such as the data sources, the uses of the app, the technologies, and protocols used in those data sources (using REST APIs for certain data and Bluetooth or Near-field communication for other data types). The usage of the use cases might be another factor (such as having a certain set of use cases for use with multiple applications). We might want to expand the presentation layer because of how certain screens are grouped to form certain isolated features and flows inside an application (such as a settings section of the application, or a login/sign-up flow). One interesting aspect to note is the :app module, which has the role of combining all of the dependencies and assembling them together. Here, we will gather all the required dependencies and initialize them.

An important thing to note here is that the modules aren't equivalent to the layers themselves; data modules can have dependencies to lower-level data modules. In fact, this situation will occur in scenarios when a module from a layer will need to have a dependency on another module from the same layer. If we were to create a dependency between the two, we might end up with a cyclical dependency, which is not wanted. In that situation, we will need to create a common module between the two that will hold the required dependencies. For example, if we want to navigate from a screen in `:presentation1` to a screen in `:presentation2` or any of the other ones, we will need to create a new module on which all of the presentation modules will depend and which will store the data or logic required to handle the navigation. We will look at this issue in more detail when we discuss the presentation layer.

To create a new Android Studio module, you need to right-click the project in Android Studio, select **New**, and then **Module**, as shown in the following figure:

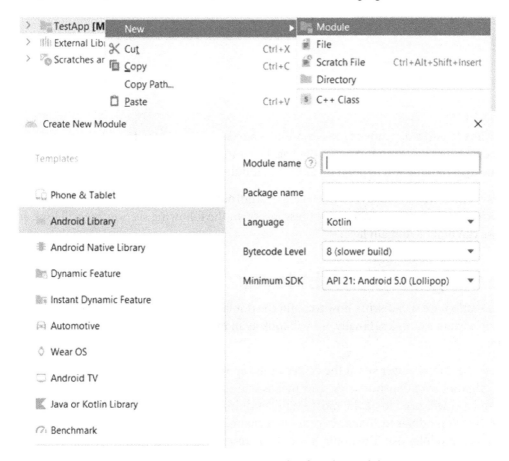

Figure 5.4 – Creating a new Android Studio module

You will then be prompted to select the type of module, and depending on the functionality, you can select **Android Library** if the module doesn't contain code from the Android framework or **Java or Kotlin Library** if the module doesn't have any dependencies to the Android framework. In the exercises that follow, we will be using Android libraries. Once the module is created, it will already contain a set of generated files and folders. One of the most relevant ones will be the `build.gradle` file. The plugin section in the file will indicate that an Android library was created:

```
plugins {
    id 'com.android.library'
    ...
}
```

If we want to add a dependency to the newly created module, we can use the following in the app module:

```
dependencies {
    implementation(project(path: ":my-new-module"))
    ...
}
```

The syntax to add a dependency to a module is similar to the syntax to add an external dependency and it's through the Gradle `implementation` method. The rest indicates that the app module will depend on another module inside the same project.

In this section, we have analyzed the layers of Android app architecture and how we can apply clean architecture principles to these layers. In the following section, we will look at how we can build a domain layer.

Creating the domain layer

In this section, we will discuss how to build the domain layer and what goes into it through certain examples. Finally, we will look at an exercise in which a domain layer is created.

Because the domain layer sits at the center of the application, it will need to have a minimal number of dependencies. This means that the Gradle modules that form the domain layer will need to be the most stable modules in the project. This is to avoid causing other modules to change because of a change that occurred in a dependency that the domain modules use. The domain should be responsible for defining the entities and use cases for the application.

Entities are represented by objects that hold data and are mainly immutable. Let's assume we want to represent a user as an entity. We might end up with something like the following:

```
data class User(
    val id: String,
    val firstName: String,
    val lastName: String,
    val email: String
) {

    fun getFullName() = "$firstName $lastName"

}
```

Here, we use a simple `data class`, and we declare all our fields immutable with the `val` keyword. We also have a business logic function for this object, which will return the full name of the user.

Next, we need to define our use cases. Because the use cases will need to get data from the data layer, we will first need to create an abstraction for our repository, and we will end up with the following:

```
interface UserRepository {

    fun getUser(id: String): User

}
```

Here, we just have a simple method that will return a user based on `id`. We can now create a use case for retrieving the user:

```
class GetUserUseCase(private val userRepository:
UserRepository) {

    fun getUser(id: String) = userRepository.getUser(id)

}
```

In the preceding example, we define a use case to retrieve the user, which will have a dependency on the `UserRepository`, which will be used to retrieve the user information. If we look at the preceding example, we can see a bit of redundancy because the use case doesn't have any extra logic and just returns the value of the repository. The benefit of use cases comes when we want to combine multiple results of multiple repositories.

Let's assume that we want to associate the user with a particular location, defined as follows:

```
data class Location(
    val id: String,
    val userId: String,
    val lat: Double,
    val long: Double
)
```

Here, we just keep the latitude and longitude associated with a particular user. Now, let's assume that we would have a repository for the different locations:

```
interface LocationRepository {

    fun getLocation(userId: String): Location
}
```

Here, we again have an abstraction of a repository with a method to get a specific location based on userId. If we want to get a user and an associated location, we will need to create a specific use case for this:

```
class GetUserWithLocationUseCase(
    private val userRepository: UserRepository,
    private val locationRepository: LocationRepository
) {

    fun getUser(id: String) =
        UserWithLocation(userRepository.getUser(id),
locationRepository.getLocation(id))
}

data class UserWithLocation(
    val user: User,
    val location: Location
)
```

In the preceding example, we create a new entity called `UserWithLocation`, which will store `User` and `Location`. `UserWithLocation` will then be used as a result for the `getUser` method in `GetUserWithLocationUseCase`. This will depend on both `UserRepository` and `LocationRepository` to fetch the relevant data.

We can further improve the use cases by handling the threading as well. Because use cases will mainly deal with retrieving and managing data, which needs to be asynchronous, we should handle this on a separate thread. We can use **Kotlin flows** to manage this, and we might end up with something like this for the repositories:

```
interface UserRepository {

    fun getUser(id: String): Flow<User>
}


interface LocationRepository {

    fun getLocation(id: String): Flow<Location>
}
```

Here, we change the return types of the methods to a Kotlin flow, which might emit a stream of data or a single item. Now, we can combine the different flows in the stream in the use case:

```
class GetUserWithLocationUseCase(
    private val userRepository: UserRepository,
    private val locationRepository: LocationRepository
) {

    fun getUser(id: String) = combine(
        userRepository.getUser(id),
        locationRepository.getLocation(id)
    ) { user, location ->
        UserWithLocation(user, location)
    }.flowOn(Dispatchers.IO)
}
```

Here, we combine the `User` and `Location` flows into a `UserWithLocation` flow, and we will execute the data fetching on the `IO` dispatcher.

Often, when dealing with data loading and management, especially from the internet, we can encounter different errors, which we will have to factor into our use cases. To solve this, we can define error entities. There are many possibilities to define them, including extending the Throwable class, defining a particular data class, a combination of the two, or combining them with sealed classes:

```kotlin
sealed class UseCaseException(override val cause: Throwable?) :
Throwable(cause) {

    class UserException(cause: Throwable) :
        UseCaseException(cause)

    class LocationException(cause: Throwable) :
        UseCaseException(cause)

    class UnknownException(cause: Throwable) :
        UseCaseException(cause)

    companion object {

        fun extractException(throwable: Throwable):
            UseCaseException {
            return if (throwable is UseCaseException)
                throwable else UnknownException(throwable)
        }
    }
}
```

Here, we have created a sealed class that will have as subclasses a dedicated error for each entity, plus an unknown error that will deal with errors we haven't accounted for, and a companion method that will check a Throwable object and return UnknownException for any Throwable that isn't UseCaseException. We will need to make sure that the error is propagated through the flow stream, but first, we can combine the entity for success with the entity for error to ensure that the consumer of the use case will not need to check the type of Throwable again and make a cast. We can do this with the following approach:

```kotlin
sealed class Result<out T : Any> {
    data class Success<out T : Any>(val data: T) :
```

```
        Result<T>()
    class Error(val exception: UseCaseException) :
        Result<Nothing>()
}
```

Here, we defined a `Result` sealed class, which will have two subclasses for success and error. The `Success` class will hold the relevant data for the use case, and the `Error` class will contain the exceptions defined before. The `Error` class can be further expanded if needed to hold data as well as the error if we want to display the cached or persisted data as a placeholder. We can now modify the use case to incorporate the `Result` class and the error state:

```
class GetUserWithLocationUseCase(
    private val userRepository: UserRepository,
    private val locationRepository: LocationRepository
) {

    fun getUser(id: String) = combine(
        userRepository.getUser(id),
        locationRepository.getLocation(id)
    ) { user, location ->
        Result.Success(UserWithLocation(user, location)) as
            Result<UserWithLocation>
    }.flowOn(Dispatchers.IO)
        .catch {
            emit(Result.Error(UseCaseException.
                extractException(it)))
        }
}
```

Here, we return `Result.Success`, which will hold the `UserWithLocation` object if no errors occur, and add use the `catch` operator to emit `Result.Error` with `UseCaseException` that occurred while fetching the data. Because these operations will repeat for multiple use cases, we can use abstraction to create a template for how each use case behaves and let the implementations deal with only processing the necessary data. An example might look like the following:

```
abstract class UseCase<T : Any, R : Any>(private val
dispatcher: CoroutineDispatcher) {
```

```
fun execute(input: T): Flow<Result<R>> =
    executeData(input)
    .map {
        Result.Success(it) as Result<R>
    }
    .flowOn(dispatcher)
    .catch {
        emit(Result.Error(UseCaseException.
            extractException(it)))
    }

internal abstract fun executeData(input: T): Flow<R>
}
```

In the preceding example, we have defined an abstract class that will contain the execute method, which will invoke the abstract executeData method and then map the result of that method into a Result object, followed by setting the flow on a CoroutineDispatcher, and finally, handling the errors in the catch operator. The implementation of this will look like the following. Note that the internal keyword for the executeData method will only make the method accessible in the current module. This is because we only want the execute method to be called by the users of this use case:

```
class GetUserWithLocationUseCase(
    dispatcher: CoroutineDispatcher,
    private val userRepository: UserRepository,
    private val locationRepository: LocationRepository
) : UseCase<String, UserWithLocation>(dispatcher) {

    override fun executeData(input: String):
        Flow<UserWithLocation> {
        return combine(
            userRepository.getUser(input),
            locationRepository.getLocation(input)
        ) { user, location ->
            UserWithLocation(user, location)
        }
```

```
        }
    }
```

In this example, `GetUserWithLocationUseCase` will only have to deal with returning the necessary data relevant to the use case in the `executeData` method. We can use generics to bind the types of data we want the use case to process by introducing further abstractions for the required input and output of it:

```
abstract class UseCase<T : UseCase.Request, R : UseCase.
Response>(private val dispatcher: CoroutineDispatcher) {

    ...

    interface Request

    interface Response
}
```

Here, we have bound the generics in the `UseCase` class to two interfaces – `Request` and `Response`. The former is represented by the input data required by the use case, and the latter is represented by the output of the use case. The implementation will now look like this:

```
class GetUserWithLocationUseCase(
    dispatcher: CoroutineDispatcher,
    private val userRepository: UserRepository,
    private val locationRepository: LocationRepository
    ) : UseCase<GetUserWithLocationUseCase.Request,
GetUserWithLocationUseCase.Response>(dispatcher) {

    override fun executeData(input: Request): Flow
        <Response> {
        return combine(
            userRepository.getUser(input.userId),
            locationRepository.getLocation(input.userId)
        ) { user, location ->
            Response(UserWithLocation(user, location))
        }
    }
}
```

```
data class Request(val userId: String) : UseCase.
    Request
data class Response(val userWithLocation:
    UserWithLocation) : UseCase.Response
}
```

Here, we provided the implementations for the `Request` and `Response` classes and used them when extending from the base class. In this case, the `Request` and `Response` classes represent **data transport objects**. When we create templates for use cases, it is important to observe their evolution because as the complexity increases, the template may become inadequate.

Often, we will have the opportunity to build a new use case from existing smaller use cases. Let's assume that for retrieving a user and retrieving a location, we have two separate use cases:

```
class GetUserUseCase(
    dispatcher: CoroutineDispatcher,
    private val userRepository: UserRepository
    ) : UseCase<GetUserUseCase.Request,
    GetUserUseCase.Response>(dispatcher) {

    override fun executeData(input: Request): Flow
        <Response> {
        return userRepository.getUser(input.userId)
            .map {
                Response(it)
            }
    }

    data class Request(val userId: String) : UseCase.
        Request
    data class Response(val user: User) : UseCase.Response
}

class GetLocationUseCase(
```

```
    dispatcher: CoroutineDispatcher,
    private val locationRepository: LocationRepository
    ) : UseCase<GetLocationUseCase.Request,
    GetLocationUseCase.Response>(dispatcher) {

    override fun executeData(input: Request): Flow
        <Response> {
        return locationRepository.getLocation(input.userId)
            .map {
                Response(it)
            }
    }

    data class Request(val userId: String) : UseCase
        .Request
    data class Response(val location: Location) : UseCase.
        Response
}
```

In the preceding examples, we have two classes for each use case to retrieve a user and a location.

We can modify GetUserWithLocationUseCase to instead use the existing use cases, like this:

```
class GetUserWithLocationUseCase(
    dispatcher: CoroutineDispatcher,
    private val getUserUseCase: GetUserUseCase,
    private val getLocationUseCase: GetLocationUseCase
    ) : UseCase<GetUserWithLocationUseCase.Request,
        GetUserWithLocationUseCase.Response>(dispatcher) {

    override fun executeData(input: Request): Flow
        <Response> {
        return combine( getUserUseCase.executeData
                (GetUserUseCase.Request(input.userId)),
            getLocationUseCase.executeData
                (GetLocationUseCase.Request(input.userId))
```

```
        ) { userResponse, locationResponse ->
        Response(UserWithLocation(userResponse.user,
            locationResponse.location))
    }
  }

  data class Request(val userId: String) : UseCase
      .Request
  data class Response(val userWithLocation:
      UserWithLocation) : UseCase.Response
}
```

Here, we changed the dependencies to instead use two existing use cases instead of the repositories, invoked the `executeData` method from each one, and then built a new `Response` using the responses from both use cases.

In this section, we looked at how to build a domain layer with entities, use cases, and abstractions for repositories. In the section that follows, we will look at an exercise related to building a domain layer.

Exercise 05.01 – Building a domain layer

In this exercise, we will create a new project in **Android Studio** in which a new Gradle module named `domain` will be created. This module will contain entities containing the following data:

- `User`: This will have an ID of the `Long` type and a name, username, and email.
- `Post`: This will have an ID and a user ID as a `Long` type, a title, and body.
- `Interaction`: This will contain the total number of interactions with the app.
- `Errors`: This is for when posts or users cannot be loaded.

The following use cases will need to be defined for the application:

- Retrieving a list containing posts with user information, grouped with the interaction data
- Retrieving information about a particular user based on the ID
- Retrieving information about a particular post based on the ID
- Updating the interaction data

To complete this exercise, you will need to do the following:

- Create a new project in Android Studio.
- Create a mapping of all the library dependencies and the versions in the root `build.gradle` file.
- Create the `domain` module in Android Studio.
- Create the required entities for the data and errors.
- Create the `Result` class, which will hold the success and error scenario.
- Create the repository abstractions to obtain the user, post, and interaction information.
- Create the four required use cases.

Follow these steps to complete the exercise:

1. Create a new project in Android Studio and select **Empty Compose Activity**.
2. In the root `build.gradle` file, add the following configurations that will be used for all the modules in the project:

```
buildscript {
    ext {
        javaCompileVersion = JavaVersion.VERSION_1_8
        jvmTarget = "1.8"
        defaultCompileSdkVersion = 31
        defaultTargetSdkVersion = 31
        defaultMinSdkVersion = 21

        ...

    }
}
```

3. In the same file, add the versions of the libraries that will be used by the Gradle modules:

```
buildscript {
    ext {

        ...

        versions = [
                androidGradlePlugin: "7.0.4",
                kotlin             : "1.5.31",
                hilt               : "2.40.5",
```

```
                        coreKtx            : "1.7.0",
                        appCompat          : "1.4.1",
                        compose            : "1.0.5",
                        lifecycleRuntimeKtx: "2.4.0",
                        activityCompose    : "1.4.0",
                        material           : "1.5.0",
                        coroutines         : "1.5.2",
                        junit              : "4.13.2",
                        mockito            : "4.0.0",
                        espressoJunit      : "1.1.3",
                        espressoCore       : "3.4.0"
            ]
        ...
    }
```

4. In the same file, add a mapping for the plugin dependencies that the entire project will use:

```
buildscript {
      ext {
            ...
            gradlePlugins = [
                        android: "com.android.tools.build:
                              gradle:${versions.
                                    androidGradlePlugin}",
                        kotlin : "org.jetbrains.kotlin:kotlin-
                              gradle-plugin:${versions.kotlin}",
                        hilt    : "com.google.dagger:hilt-
                              android-gradle-plugin:
                                    ${versions.hilt}"
            ]
            ...
      }
```

5. Next, you will need to add the dependencies to the androidx libraries:

```
buildscript {
      ext {
```

```
         ...
        androidx = [
                core                    : "androidx.
core:core-ktx:${versions.coreKtx}",
                appCompat               : "androidx.
appcompat:appcompat:${versions.appCompat}",
                composeUi               : "androidx.
compose.ui:ui:${versions.compose}",
                composeMaterial         : "androidx.
compose.material:material:${versions.compose}",
                composeUiToolingPreview: "androidx.
compose.ui:ui-tooling-preview:${versions.compose}",
                lifecycleRuntimeKtx     : "androidx.
lifecycle:lifecycle-runtime-ktx:${versions.
lifecycleRuntimeKtx}",
                composeActivity         : "androidx.
activity:activity-compose:${versions.activityCompose}"
        ]
        ...
}
```

6. Next, add the remaining libraries for material design, dependency injection, and tests:

```
buildscript {
    ext {
        ...
        material = [
                material: "com.google.android.
                    material:material:$
                        {versions.material}"
        ]
        coroutines = [
                coroutinesAndroid: "org.jetbrains.
                    kotlinx:kotlinx-coroutines-
                        android:${versions.coroutines}"
        ]
        di = [
                hiltAndroid : "com.google.dagger:hilt-
```

```
                    android:${versions.hilt}",
                hiltCompiler: "com.google.dagger:hilt-
                    compiler:${versions.hilt}"
        ]
        test = [
                junit     :
                    "junit:junit:${versions.junit}",
                coroutines: "org.jetbrains.kotlinx:
                    kotlinx-coroutines-test:
                        ${versions.coroutines}",
                mockito    : "org.mockito.kotlin:
                    mockito-kotlin:${versions.mockito}"
        ]
        androidTest = [
                junit              : "androidx.test.ext
                    :junit:${versions.espressoJunit}",
                espressoCore       : "androidx.test.
                    espresso:espresso-core:$
                        {versions.espressoCore}",
                composeUiTestJunit: "androidx.compose.
                    ui:ui-test-junit4:${versions.compose}"
        ]
    }
    ...
}
```

7. In the same file, you will need to replace the previous mappings as plugin
    dependencies:

```
buildscript {
    ...
    dependencies {
        classpath gradlePlugins.android
        classpath gradlePlugins.kotlin
        classpath gradlePlugins.hilt
    }
}
```

8. Now, you need to switch to the `build.gradle` file in the app module and change the existing configurations with the ones defined in the top-level `build.gradle`:

```
android {
    compileSdk defaultCompileSdkVersion
    defaultConfig {
        ...
        minSdk defaultMinSdkVersion
        targetSdk defaultTargetSdkVersion
        versionCode 1
        versionName "1.0"
        ...
    }
    ...
    compileOptions {
        sourceCompatibility javaCompileVersion
        targetCompatibility javaCompileVersion
    }
    kotlinOptions {
        jvmTarget = jvmTarget
        useIR = true
    }
    buildFeatures {
        compose true
    }
    composeOptions {
        kotlinCompilerExtensionVersion
            versions.compose
    }
    ...
}
```

9. In the same file, you will need to replace the dependencies with the ones defined in the top-level `build.gradle` file:

```
dependencies {

    implementation androidx.core
```

```
implementation androidx.appCompat
implementation material.material
implementation androidx.composeUi
implementation androidx.composeMaterial
implementation androidx.composeUiToolingPreview
implementation androidx.lifecycleRuntimeKtx
implementation androidx.composeActivity
testImplementation test.junit
}
```

10. In Android Studio, execute **Sync Project with Gradle Files** command and then the **Make Project** command to make sure that the project builds without any errors.

11. Create a new module for the project named domain, which will be an Android library module.

12. In the build.gradle file of the domain module, make sure you have the following plugins:

```
plugins {
    id 'com.android.library'
    id 'kotlin-android'
    id 'kotlin-kapt'
    id 'dagger.hilt.android.plugin'
}
```

13. In the same file, make sure you use the configurations defined in the top-level build.gradle file:

```
android {
    compileSdk defaultCompileSdkVersion

    defaultConfig {
        minSdk defaultMinSdkVersion
        targetSdk defaultTargetSdkVersion

        ...
    }
    ...
    compileOptions {
        sourceCompatibility javaCompileVersion
        targetCompatibility javaCompileVersion
```

```
    }
    kotlinOptions {
        jvmTarget = jvmTarget
    }
}
```

14. In the same file, you will need to add the following dependencies:

```
dependencies {
    implementation coroutines.coroutinesAndroid
    implementation di.hiltAndroid
    kapt di.hiltCompiler
    testImplementation test.junit
    testImplementation test.coroutines
    testImplementation test.mockito
}
```

15. Sync the project with Gradle files and build the project again to make sure that the Gradle configuration is correct.

16. In the domain module, create a new package named entity.

17. In the entity package, create a class named Post, which will have id, userId, title, and body:

```
data class Post(
    val id: Long,
    val userId: Long,
    val title: String,
    val body: String
)
```

18. In the same package, create the User class, which will have id, name, username, and email:

```
data class User(
    val id: Long,
    val name: String,
    val username: String,
    val email: String
)
```

19. Next, create a class named `PostWithUser`, which will contain the `post` and `user` information:

```
data class PostWithUser(
    val post: Post,
    val user: User
)
```

20. In the same package, create a class called `Interaction`, which will contain the total number of clicks:

```
data class Interaction(val totalClicks: Int)
```

21. Now, we need to create the error entities:

```
sealed class UseCaseException(cause: Throwable) :
Throwable(cause) {

    class PostException(cause: Throwable) :
        UseCaseException(cause)

    class UserException(cause: Throwable) :
        UseCaseException(cause)

    class UnknownException(cause: Throwable) :
        UseCaseException(cause)


    companion object {

        fun createFromThrowable(throwable: Throwable):
            UseCaseException {
            return if (throwable is UseCaseException)
            throwable else UnknownException(throwable)
        }
    }
}
```

Here, we have exceptions defined for when there will be an issue with loading the post and user information, and `UnknownException`, which will be emitted when something else goes wrong.

22. Next, let's create the `Result` class, which will hold the success and error information:

```
sealed class Result<out T : Any> {
    data class Success<out T : Any>(val data: T) :
        Result<T>()
    class Error(val exception: UseCaseException) :
        Result<Nothing>()
}
```

23. Now, we need to move on to defining the abstractions for the repositories, and to do so, we create a new package named `repository`.

24. In the `repository` package, create an interface for managing the post data:

```
interface PostRepository {

    fun getPosts(): Flow<List<Post>>

    fun getPost(id: Long): Flow<Post>
}
```

25. In the same package, create an interface for managing the user data:

```
interface UserRepository {

    fun getUsers(): Flow<List<User>>

    fun getUser(id: Long): Flow<User>
}
```

26. In the same package, create an interface for managing the interaction data:

```
interface InteractionRepository {

    fun getInteraction(): Flow<Interaction>
```

```
        fun saveInteraction(interaction: Interaction):
            Flow<Interaction>
}
```

27. Now, we move on to the use cases and start by creating a new package named usecase.

28. In this package, create the UseCase template:

```
abstract class UseCase<I : UseCase.Request, O : UseCase.
Response>(private val configuration: Configuration) {

    fun execute(request: I) = process(request)
        .map {
            Result.Success(it) as Result<O>
        }
        .flowOn(configuration.dispatcher)
        .catch {
            emit(Result.Error(UseCaseException.
                createFromThrowable(it)))
        }

    internal abstract fun process(request: I): Flow<O>

    class Configuration(val dispatcher:
        CoroutineDispatcher)

    interface Request

    interface Response
}
```

In this template, we have defined the data transfer objects' abstraction, as well as a Configuration class that holds CoroutineDispatcher. The reason for this Configuration class is to be able to add other parameters for the use case without modifying the UseCase subclasses. We have one abstract method, which will be implemented by the subclasses to retrieve the data from the repositories, and the execute method, which will take the data and convert it to Result, handle the error scenarios, and set the proper CoroutineDispatcher.

29. In the `usecase` package, create the use case to retrieve the list of posts with the user information and the interaction data:

```
class GetPostsWithUsersWithInteractionUseCase @Inject
constructor(
    configuration: Configuration,
    private val postRepository: PostRepository,
    private val userRepository: UserRepository,
    private val interactionRepository:
        InteractionRepository
) : GetPostsWithUsersWithInteractionUseCase
GetPostsWithUsersWithInteractionUseCase {

    override fun process(request: Request):
        Flow<Response> =
        combine(
            postRepository.getPosts(),
            userRepository.getUsers(),
            interactionRepository.getInteraction()
        ) { posts, users, interaction ->
            val postUsers = posts.map { post ->
                val user = users.first {
                    it.id == post.userId
                }
                PostWithUser(post, user)
            }
            Response(postUsers, interaction)
        }

    object Request : UseCase.Request

    data class Response(
        val posts: List<PostWithUser>,
        val interaction: Interaction
    ) : UseCase.Response
}
```

In this class, we extend the `UseCase` class, and in the `process` method, we combine the posts, users, and interaction flows. Because there is no input required, the `Request` class will have to be empty, and the `Response` class will contain a list of combined user and post information as well as the interaction data. The `@Inject` annotation will help us inject this use case later in the presentation layer.

30. In the same package, create the use case to retrieve a post by ID:

```kotlin
class GetPostUseCase @Inject constructor(
    configuration: Configuration,
    private val postRepository: PostRepository
) : UseCase<GetPostUseCase.Request,
GetPostUseCase.Response>(configuration) {

    override fun process(request: Request): Flow
        <Response> =
        postRepository.getPost(request.postId)
            .map {
                Response(it)
            }

    data class Request(val postId: Long) : UseCase.
        Request
    data class Response(val post: Post) : UseCase.
        Response
}
```

31. In the same package, create the use case to retrieve a user by ID:

```kotlin
class GetUserUseCase @Inject constructor(
    configuration: Configuration,
    private val userRepository: UserRepository
) : UseCase<GetUserUseCase.Request,
GetUserUseCase.Response>(configuration) {

    override fun process(request: Request): Flow
        <Response> =
        userRepository.getUser(request.userId)
            .map {
```

```
                    Response(it)
            }

    data class Request(val userId: Long) : UseCase.
        Request
    data class Response(val user: User) : UseCase.
        Response
}
```

32. Now, we move on to the last use case for updating the interaction data:

```
class UpdateInteractionUseCase @Inject constructor(
    configuration: Configuration,
    private val interactionRepository:
        InteractionRepository
) : UseCase<UpdateInteractionUseCase.Request,
    UpdateInteractionUseCase.Response>(configuration) {

    override fun process(request: Request): Flow
        <Response> {
        return interactionRepository.saveInteraction
            (request.interaction)
            .map {
                Response
            }
    }

    data class Request(val interaction: Interaction) :
        UseCase.Request

    object Response : UseCase.Response
}
```

33. To unit-test the code, we need to create a new folder called resources in the test folder of the domain module.

34. Inside the `resources` folder, create a subfolder called `mockito-extensions`; inside this folder, create a file named `org.mockito.plugins.MockMaker`; and inside this file, add the following text – `mock-maker-inline`. This allows the Mockito testing library to mock the `final` Java class, which in Kotlin means all classes without the `open` keyword.

35. Create a new class named `UseCaseTest` in the test folder of the `domain` module:

```
class UseCaseTest {

    @ExperimentalCoroutinesApi
    private val configuration = UseCase.Configuration
        (TestCoroutineDispatcher())
    private val request = mock<UseCase.Request>()
    private val response = mock<UseCase.Response>()

    @ExperimentalCoroutinesApi
    private lateinit var useCase:
        UseCase<UseCase.Request, UseCase.Response>

    @ExperimentalCoroutinesApi
    @Before
    fun setUp() {
        useCase = object : UseCase<UseCase.Request,
            UseCase.Response>(configuration) {
            override fun process(request: Request):
                Flow<Response> {
                assertEquals(this@UseCaseTest.request,
                    request)
                return flowOf(response)
            }
        }
    }
}
```

Here, we provide an implementation for the `UseCase` class, which will return a mocked response.

36. Next, create a test method that will verify the successful scenario for the `execute` method:

```
@ExperimentalCoroutinesApi
@Test
fun testExecuteSuccess() = runBlockingTest {
    val result = useCase.execute(request).first()
    assertEquals(Result.Success(response), result)
}
```

Here, we assert that the result of the `execute` method is `Success` and that it contains the mocked response.

37. Next, create a new test class named `GetPostsWithUsersWithInteractionUseCaseTest`:

```
class GetPostsWithUsersWithInteractionUseCaseTest {

    private val postRepository = mock<PostRepository>()
    private val userRepository = mock<UserRepository>()
    private val interactionRepository =
mock<InteractionRepository>()
    private val useCase =
GetPostsWithUsersWithInteractionUseCase(
        mock(),
        postRepository,
        userRepository,
        interactionRepository
    )
}
```

Here, we mock all the repositories and inject the mocks into the class we want to test.

38. Finally, create a test method that will verify the `process` method from the use case we are testing:

```
@ExperimentalCoroutinesApi
@Test
fun testProcess() = runBlockingTest {
    val user1 = User(1L, "name1", "username1",
"email1")
    val user2 = User(2L, "name2", "username2",
```

```
  "email2")
        val post1 = Post(1L, user1.id, "title1", "body1")
        val post2 = Post(2L, user1.id, "title2", "body2")
        val post3 = Post(3L, user2.id, "title3", "body3")
        val post4 = Post(4L, user2.id, "title4", "body4")
        val interaction = Interaction(10)
        whenever(userRepository.getUsers()).thenReturn
            (flowOf(listOf(user1, user2)))
        whenever(postRepository.getPosts()).thenReturn
            (flowOf(listOf(post1, post2, post3, post4)))
whenever(interactionRepository.getInteraction
            ()).thenReturn(flowOf(interaction))
        val response = useCase.process
            (GetPostsWithUsersWithInteractionUseCase.
                Request).first()
        assertEquals(
            GetPostsWithUsersWithInteractionUseCase.
                Response(
                listOf(
                    PostWithUser(post1, user1),
                    PostWithUser(post2, user1),
                    PostWithUser(post3, user2),
                    PostWithUser(post4, user2),
                ), interaction
            ),
            response
        )
    }
```

Here, we provide mock lists of users and posts and a mock interaction, then we return these for each of the repository calls, and then we assert that the result is a list of four posts, written by two users and the mock interaction.

If we run the tests for these two methods, they should pass. To test the remaining use cases, the same principles can be applied that we used for GetPostsWithUsersWith-InteractionUseCaseTest – create mock repositories, inject them into the object we wish to test, and then define the mocks for the input of the process method and the results we should expect, which will give us output as shown in the following screenshot:

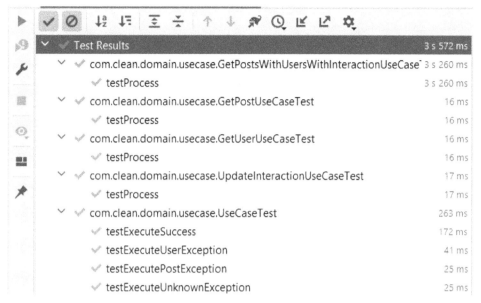

Figure 5.5 – Output of the use case unit tests

In this section, we performed an exercise in which we created a simple domain that contained entities, a few simple use cases, and a particular use case that combined multiple data sources. The domain module has dependencies on flows and Hilt. This means that changes to these libraries might cause changes to our domain module. This decision was made because of the benefits that these libraries provide when it comes to reactive programming and dependency injection. Because we considered dependency injection when defining the use cases, this made them more testable, as we could inject mock objects into the tested objects very easily.

Summary

In this chapter, we looked at how the architecture of an Android app is layered and focused on the domain layer, discussing the topics of entities and use cases. We also learned how to use dependency inversion to place use cases and entities at the center of our architecture. We did this by creating repository abstractions that can be implemented in the lower layers. We also learned how to use library modules to enforce separations between layers.

In the chapter's exercise, we created a domain module for an Android application, providing an example of what a domain layer might look like. In the next chapter, we will focus on the data layer, in which we will provide implementations for the repository abstractions we defined in the domain layer, and discuss how we can use these repositories to manage an application's data.

6
Assembling a Repository

In this chapter, we will begin by discussing the application's data layer and the components that make up this layer, including repositories and data sources. We will then move on to the topic of repositories, one of the application layer's components, and the role they play in managing the data of an application. In this chapter's exercise, we will continue the project started in the previous chapter by providing the repository implementations for the abstractions defined there and also introducing new abstractions for the different types of data sources.

In this chapter, we will cover the following topics:

- Creating the data layer
- Creating repositories

By the end of the chapter, you will have learned what the data layer is and how we can create repositories for an Android application.

Technical requirements

The hardware and software requirements are as follows:

- Android Studio Arctic Fox 2020.3.1 Patch 3

The code files for this chapter can be found here: `https://github.com/ PacktPublishing/Clean-Android-Architecture/tree/main/Chapter6`.

Check out the following video to see the Code in Action: `https://bit.ly/3NpAhNs`

Creating the data layer

In this section, we will look at the data layer of an Android application and the components that typically form part of the data layer.

The data layer is the layer in which data is created and managed. This means that this layer is responsible for creating, reading, updating, and deleting data, as well as for managing and ensuring that data from the internet is synced with persistent data.

In the previous chapter, we have seen that use cases depend on an abstraction of a repository class, and there can be multiple repositories for different data types. Repositories represent the entry point into the data layer and are responsible for managing multiple data sources and centralizing the data. The data sources represent the other component of the data layer and are responsible for managing the data of a particular source (internet, Room, data store, and suchlike).

An example of what the data layer for a particular set of data, which uses two data sources, might look like is shown in the following figure:

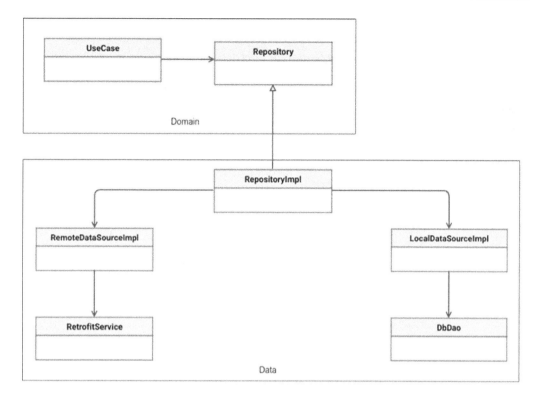

Figure 6.1 – Data layer example

In the preceding diagram, we have an example of what a data layer would look like when it is connected to the domain layer. We can observe that the UseCase class depends on a Repository abstraction, which represents the domain layer. The data layer is represented by RepositoryImpl, which is the implementation of the Repository abstraction. The RepositoryImpl class depends on the two data source implementations: RemoteDataSourceImpl and LocalDataSourceImpl. Each data source then depends on a particular implementation for managing data from the internet using Retrofit in the case of RetrofitService, or using a particular data access class that uses Room in the case of DbDao.

This approach poses a problem owing to the direct dependency between `RepositoryImpl` and `RemoteDataSourceImpl`, and the problem arises when we might want to swap out Retrofit or Room for alternatives. If we might want to swap out these libraries for others, we risk changes in the `RepositoryImpl` class, which violates the single-responsibility principle. The solution for this is like the solution we had for solving the dependencies between the use cases and the repositories, and that is to invert the dependencies between the repository and the data sources. This would look like the following:

Figure 6.2 – Data layer with inverted dependencies

In the preceding diagram, we have introduced two abstractions for each data source, named `RemoteDataSource` and `LocalDataSource`. `RepositoryImpl` now depends on these two abstractions and all the conversions between Retrofit- or Room-related objects and domain entities should now be placed in `RemoteDataSourceImpl` or `LocalDataSourceImpl`, which inherit the new abstractions and will continue to handle the data from Retrofit or Room. If we want to split the data layer into different Gradle modules, we will have the following:

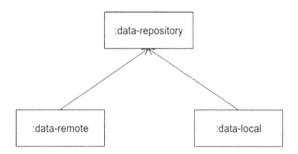

Figure 6.3 – Data layer modules

The preceding diagram shows the Gradle module dependencies between the repository and local and remote data sources. Here we can see the benefit of dependency inversion, which allows us to have a separate repository module without depending on Retrofit or Room.

In this section, we have discussed the data layer and the components inside it and how to manage the dependencies between all the components. In the following section, we will take a closer look at repositories and how to implement them.

Creating repositories

In this section, we will look at what a repository is and the role it plays in the data layer of an application, and how we can create repositories with various data sources.

The repository represents an abstraction for the data than an application uses, and it is responsible for managing and centralizing the data from one or multiple data sources.

In the previous chapter, we defined the following entity:

```
data class User(
    val id: String,
    val firstName: String,
    val lastName: String,
    val email: String
) {

    fun getFullName() = "$firstName $lastName"
}
```

Here we have a simple `User` data class with a few relevant fields. The repository abstraction for the `User` data is as follows:

```
interface UserRepository {

    fun getUser(id: String): Flow<User>
}
```

Here we have an interface named `UserRepository` that is responsible for fetching the user information in a Kotlin flow.

If we want to fetch data from the internet, we must first define a `UserRemoteDataSource` abstraction:

```
interface UserRemoteDataSource {

    fun getUser(id: String): Flow<User>
}
```

In this case, we have an interface similar to how `UserRepository` is defined with a simple method to retrieve a `User` object. We can now implement `UserRepository` to use this data source:

```
class UserRepositoryImpl(private val userRemoteDataSource:
    UserRemoteDataSource) : UserRepository {

    override fun getUser(id: String): Flow<User> =
        userRemoteDataSource.getUser(id)

}
```

Here we have a dependency on `UserRemoteDataSource` and invoke the `getUser` method. If we want to persist the remote user data locally, we will need to define a `UserLocalDataSource` abstraction, which will be responsible for inserting the user:

```
interface UserLocalDataSource {

    suspend fun insertUser(user: User)
}
```

Here we have a method for inserting a user into the local store. We can now update `UserRepositoryImpl` to connect the data sources and insert a user after it was retrieved:

```
class UserRepositoryImpl(
    private val userRemoteDataSource: UserRemoteDataSource,
    private val userLocalDataSource: UserLocalDataSource
) : UserRepository {

    override fun getUser(id: String): Flow<User> =
        userRemoteDataSource.getUser(id)
        .onEach {
            userLocalDataSource.insertUser(it)
        }

}
```

This represents a simple use case for data sources, but we can use repositories to improve the user experience for the user. For instance, we can change the repository implementation to return the saved data and have a separate method for fetching the data remotely. We can take advantage of flows, which can emit multiple users in a stream:

```
interface UserLocalDataSource {

    suspend fun insertUser(user: User)

    fun getUser(id: String): Flow<User>
}
```

In the preceding example, we have added the `getUser` method to retrieve a `User` object, which was persisted locally. We will need to modify the repository abstraction as follows:

```
interface UserRepository {

    fun getUser(id: String): Flow<User>

    fun refreshUser(id: String): Flow<User>
}
```

Here, we have added the `refreshUser` method, which, when implemented, will be responsible for fetching a new user from the internet. The implementation will be as follows:

```
class UserRepositoryImpl(
    private val userRemoteDataSource: UserRemoteDataSource,
    private val userLocalDataSource: UserLocalDataSource
) : UserRepository {

    override fun getUser(id: String): Flow<User> =
        userLocalDataSource.getUser(id)

    override fun refreshUser(id: String): Flow<User> =
        userRemoteDataSource.getUser(id)
        .onEach {
            userLocalDataSource.insertUser(it)
        }

}
```

Here, we return the persisted user in the `getUser` method and, in the `refreshUser` method, we now fetch the remote data and insert it locally. If we are using libraries such as Room, this will trigger the emission of a new `User` object, which will come from `UserLocalDataSource`. This means that all subscribers of the `getUser` method will be notified of a change and receive a new `User` object.

We can also use repositories for caching data in the memory. An example of this would be as follows:

```
class UserRepositoryImpl(
    private val userRemoteDataSource: UserRemoteDataSource,
    private val userLocalDataSource: UserLocalDataSource
) : UserRepository {

    private val usersFlow = MutableStateFlow
        (emptyMap<String, User>().toMutableMap())

    override fun getUser(id: String): Flow<User> =
        usersFlow.flatMapLatest {
```

```
            val user = it[id]
            if (user != null) {
                flowOf(user)
            } else {
                userLocalDataSource.getUser(id)
                    .onEach { persistedUser ->
                        saveUser(persistedUser)
                    }
            }
        }

    override fun refreshUser(id: String): Flow<User> =
        userRemoteDataSource.getUser(id)
        .onEach {
            saveUser(it)
            userLocalDataSource.insertUser(it)
        }

    private fun saveUser(user: User) {
        val map = usersFlow.value
        map[user.id] = user
        usersFlow.value = map
    }

}
```

Here, we have added a new `MutableStateFlow` object, which will hold a map in which the keys are represented by the user IDs and the values are the users. In the `getUser` method, we check whether the user is stored in memory and return the memory value if present, otherwise we get the persisted data, which we will store in memory after. In the `refreshUser` method, we persist the value in memory and persist the data locally.

Because we defined the repository abstraction to return entities, we should try as much as possible to use entities across the repository and the data source abstractions. However, we might need specific object definitions to handle processing the data from the data sources. We can define these specific classes in this layer and then convert them to entities in the repository implementation.

In this section, we have seen how we can create repositories and how they can be used to manage data in an application. In the section that follows, we will look at an exercise in which we will create repositories for an application.

Exercise 06.01 – Creating repositories

Modify *Exercise 05.01: Building a domain layer*, so that a new library module is created in Android Studio. The module will be named `data-repository` and will have a dependency on the `domain` module. In this module, we will implement the repository classes from the domain module as follows:

- `UserRepositoryImpl` will have dependencies on the following data sources: `UserRemoteDataSource`, which will fetch a list and a user by ID, and `UserLocalDataSource`, which will have methods for inserting a list of users and obtaining a list of the same. `UserRepositoryImpl` will always load the remote users and insert them locally.

- `PostRepositoryImpl` will have dependencies on the following data sources: `PostRemoteDataSource`, which will fetch a list of users and a user by ID, and `PostLocalDataSource`, which will have methods for inserting a list of posts and obtaining a list of the same. `PostRepositoryImpl` will always load the remote posts and insert them locally.

- `InteractionRepositoryImpl` will have a dependency on a single data source, `LocalInteractionDataSource`, which will be responsible for loading an interaction and saving it. `InteractionRepositoryImpl` will load the interaction and save a new interaction.

To complete this exercise, you will need to do the following:

- Create the data repository module in Android Studio
- Create the user's data sources and repository
- Create the post's data sources and repository
- Create the interaction data source and repository

Follow these steps to complete the exercise:

1. Create a new module named `data-repository`, which will be an Android Library module.

2. Make sure that in the top-level `build.gradle` file, the following dependencies are set:

```
buildscript {
```

```
    ...
    dependencies {
        classpath gradlePlugins.android
        classpath gradlePlugins.kotlin
        classpath gradlePlugins.hilt
    }
}
```

3. In the `build.gradle` file of the `data-repository` module, make sure that the following plugins are present:

```
plugins {
    id 'com.android.library'
    id 'kotlin-android'
    id 'kotlin-kapt'
    id 'dagger.hilt.android.plugin'
}
```

4. In the same file, change the configurations to the ones defined in the top-level `build.gradle` file:

```
android {
    compileSdk defaultCompileSdkVersion

    defaultConfig {
        minSdk defaultMinSdkVersion
        targetSdk defaultTargetSdkVersion
        ...
    }
    ...
    compileOptions {
        sourceCompatibility javaCompileVersion
        targetCompatibility javaCompileVersion
    }
    kotlinOptions {
        jvmTarget = jvmTarget
    }
}
```

5. In the same file, make sure that the following dependencies are specified:

```
dependencies {
    implementation(project(path: ":domain"))
    implementation coroutines.coroutinesAndroid
    implementation di.hiltAndroid
    kapt di.hiltCompiler
    testImplementation test.junit
    testImplementation test.coroutines
    testImplementation test.mockito
}
```

Here, we are using the implementation method to add a dependency to the :domain module, in the same way as other libraries are referenced. In Gradle we also have the option of using the api method. This makes a module's dependencies public to other modules. This, in turn, might have potential side effects, such as leaking dependencies that should be kept private. In this example, we might be better served by using the api method for the :domain module because of the close relationship between the two modules (which would make all modules that depend on :data-repository not have to add the dependency to :domain). However, dependencies such as Hilt and Coroutines should be kept with the implementation method because we would want to avoid exposing these libraries in modules that do not use them.

6. In the data-repository module, create a new package named data_source.

7. In the data_source package, create a new package named remote.

8. In the remote package, create the RemoteUserDataSource interface:

```
interface RemoteUserDataSource {

    fun getUsers(): Flow<List<User>>

    fun getUser(id: Long): Flow<User>
}
```

9. In the remote package, create the RemotePostDataSource interface:

```
interface RemotePostDataSource {

    fun getPosts(): Flow<List<Post>>
```

```
    fun getPost(id: Long): Flow<Post>
}
```

10. In the data_source package, create a new package called local.

11. In the local package, create the LocalUserDataSource interface:

```
interface LocalUserDataSource {

    fun getUsers(): Flow<List<User>>

    suspend fun addUsers(users: List<User>)
}
```

12. In the local package, create the LocalPostDataSource interface:

```
interface LocalPostDataSource {

    fun getPosts(): Flow<List<Post>>

    suspend fun addPosts(posts: List<Post>)
}
```

13. In the local package, create the LocalInteractionDataSource package:

```
interface LocalInteractionDataSource {

    fun getInteraction(): Flow<Interaction>

    suspend fun saveInteraction(interaction: Interaction)
}
```

14. Next to the data_source package, create a new package named repository.

15. In the repository package, create the UserRepositoryImpl class:

```
class UserRepositoryImpl @Inject constructor(
    private val remoteUserDataSource:
        RemoteUserDataSource,
    private val localUserDataSource:
        LocalUserDataSource
```

```kotlin
) : UserRepository {

    override fun getUsers(): Flow<List<User>> =
        remoteUserDataSource.getUsers()
        .onEach {
            localUserDataSource.addUsers(it)
        }

    override fun getUser(id: Long): Flow<User> =
remoteUserDataSource.getUser(id)
        .onEach {
            localUserDataSource.addUsers(listOf(it))
        }

}
```

Here, we fetch the user data from the remote data source and store it locally.

16. In the same package, create the `PostRepositoryImpl` class:

```kotlin
class PostRepositoryImpl @Inject constructor(
    private val remotePostDataSource:
        RemotePostDataSource,
    private val localPostDataSource:
        LocalPostDataSource
) : PostRepository {

    override fun getPosts(): Flow<List<Post>> =
        remotePostDataSource.getPosts()
        .onEach {
            localPostDataSource.addPosts(it)
        }

    override fun getPost(id: Long): Flow<Post> =
        remotePostDataSource.getPost(id)
        .onEach {
            localPostDataSource.addPosts(listOf(it))
        }

}
```

Here, we are fetching the post data from the remote data source and using the local data source to persist the data.

17. In the same package, create the `InteractionRepositoryImpl` class:

```
class InteractionRepositoryImpl @Inject constructor(
    private val interactionDataSource:
        LocalInteractionDataSource
) : InteractionRepository {

    override fun getInteraction(): Flow<Interaction> =
        interactionDataSource.getInteraction()

    override fun saveInteraction(interaction:
        Interaction): Flow<Interaction> = flow {
        interactionDataSource.saveInteraction(interaction)
        this.emit(Unit)
    }.flatMapLatest {
        getInteraction()
    }
}
```

Here, we are just interacting with the local data source to read and store the data.

18. We now want to use Hilt to bind the repository abstraction with the implementation, so we will need to create a package named `injection` next to the `data_source` and `repository` packages.

19. Inside the `injection` package, create a class named `RepositoryModule`:

```
@Module
@InstallIn(SingletonComponent::class)
abstract class RepositoryModule {

    @Binds
    abstract fun bindPostRepository(postRepositoryImpl
        : PostRepositoryImpl): PostRepository

    @Binds
    abstract fun bindUserRepository
        (userRepositoryImpl: UserRepositoryImpl):
```

```
        UserRepository

    @Binds
    abstract fun bindInteractionRepository
        (interactionRepositoryImpl:
            InteractionRepositoryImpl):
                InteractionRepository
}
```

Here, we are using the `@Binds` Hilt annotation, which maps the implementation of a repository annotated with `@Inject` with the abstraction.

20. To unit test the code, we will now need to create a new folder called `resources` in the test folder of the `data-repository` module.

21. Inside the resources folder, create a folder called `mockito-extensions` and, inside this folder, create a file named `org.mockito.plugins.MockMaker`, and, inside this file, add the following text: `mock-maker-inline`.

22. Create a `UserRepositoryImplTest` class for unit testing the `UserRepositoryImpl` methods:

```
class UserRepositoryImplTest {

    private val remoteUserDataSource =
        mock<RemoteUserDataSource>()
    private val localUserDataSource =
        mock<LocalUserDataSource>()
    private val repositoryImpl = UserRepositoryImpl
        (remoteUserDataSource, localUserDataSource)

}
```

23. In the `UserRepositoryImplTest` class, add a test method for each repository method:

```
class UserRepositoryImplTest {
    ...
    @ExperimentalCoroutinesApi
    @Test
    fun testGetUsers() = runBlockingTest {
```

```
        val users = listOf(User(1, "name", "username",
            "email"))
        whenever(remoteUserDataSource.getUsers()).
            thenReturn(flowOf(users))
        val result = repositoryImpl.getUsers().first()
        assertEquals(users, result)
        verify(localUserDataSource).addUsers(users)
    }


    @ExperimentalCoroutinesApi
    @Test
    fun testGetUser() = runBlockingTest {
        val id = 1L
        val user = User(id, "name", "username",
            "email"
)
        whenever(remoteUserDataSource.getUser(id))
            .thenReturn(flowOf(user))
        val result = repositoryImpl.getUser(id).
            first()
        assertEquals(user, result)
        verify(localUserDataSource).
addUsers(listOf(user))
    }
}
```

In this class, we unit test each of the methods in the `UserRepositoryImpl` class by mocking the local data and remote data sources and verifying that the data obtained from the remote data source is inserted into the local data source.

24. Create a `PostRepositoryImplTest` class to test the `PostRepositoryImpl` class:

```
class PostRepositoryImplTest {

    private val remotePostDataSource =
        mock<RemotePostDataSource>()
    private val localPostDataSource =
```

```
                    mock<LocalPostDataSource>()
        private val repositoryImpl = PostRepositoryImpl
            (remotePostDataSource, localPostDataSource)
}
```

25. Create unit tests for each of the methods in the `PostRepositoryImpl` class:

```
class PostRepositoryImplTest {
    ...
    @ExperimentalCoroutinesApi
    @Test
    fun testGetPosts() = runBlockingTest {
        val posts = listOf(Post(1, 1, "title",
            "body"))
        whenever(remotePostDataSource.getPosts())
            .thenReturn(flowOf(posts))
        val result = repositoryImpl.getPosts().first()
        Assert.assertEquals(posts, result)
        verify(localPostDataSource).addPosts(posts)
    }

    @ExperimentalCoroutinesApi
    @Test
    fun testGetPost() = runBlockingTest {
        val id = 1L
        val post = Post(id, 1, "title", "body")
        whenever(remotePostDataSource.getPost(id)).
thenReturn(flowOf(post))
        val result =
            repositoryImpl.getPost(id).first()
        Assert.assertEquals(post, result)
        verify(localPostDataSource).
addPosts(listOf(post))
    }
}
```

In this class, we perform the same tests that we did for `UserRepositoryImpl`.

26. Create an `InteractionRepositoryImplTest` class to test the `InteractionRepositoryImpl` class:

```
class InteractionRepositoryImplTest {

    private val localInteractionDataSource =
        mock<LocalInteractionDataSource>()
    private val repositoryImpl =
        InteractionRepositoryImpl
        (localInteractionDataSource)
}
```

27. Create unit tests for each of the methods in the `InteractionRepositoryImpl` class:

```
class InteractionRepositoryImplTest {
    ...
    @ExperimentalCoroutinesApi
    @Test
    fun testGetInteraction() = runBlockingTest {
        val interaction = Interaction(10)
        whenever(localInteractionDataSource.
            getInteraction()).
                thenReturn(flowOf(interaction))
        val result = repositoryImpl.getInteraction()
            .first()
        assertEquals(interaction, result)
    }

    @ExperimentalCoroutinesApi
    @Test
    fun testSaveInteraction() = runBlockingTest {
        val interaction = Interaction(10)
        whenever(localInteractionDataSource.
            getInteraction()).thenReturn
                (flowOf(interaction))
        val result = repositoryImpl.saveInteraction
            (interaction).first()
```

```
            veriy(localInteractionDataSource).
                saveInteraction(interaction)
        assertEquals(interaction, result)
        }
    }
```

In this class, we mock the local data source and then we verify that the repository has the appropriate invocations on the `LocalInteractionDataStore` mock.

If we run the tests, we should see something like the following screenshot:

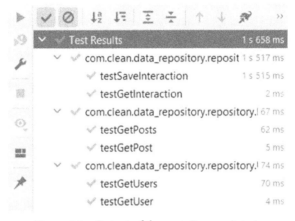

Figure 6.4 – Output of the repository unit tests

In this exercise, we have created a new module in which we implemented our repositories and defined new abstractions for the data sources that the repositories will use. Here, we have continued the integration with other libraries, such as Hilt for dependency injection, and Kotlin flows to handle the data in a reactive approach. The use of dependency injection made the unit tests simple to write because we could easily provide mocks.

Summary

In this chapter, we started looking into the data layer of an Android application and provided an overview of the components that are part of this layer. We also looked at the Repository component, which is responsible for managing the data provided by one or more data sources, and provided examples of how we could build different repositories. We also looked at the relationship between repositories and data sources and how we can further decouple the components with dependency inversion, to keep our repositories unaffected by changes in libraries used to fetch data. Finally, we looked at an exercise on how we can build repositories with local and remote data sources. In the following chapter, we will continue with the data layer and how we can integrate the remote and local data sources with libraries such as Room and Retrofit.

7
Building Data Sources

In this chapter, we will continue focusing on the data layer by discussing how we can implement local and remote data sources and the roles they play in clean architecture. First, we will look at how remote data sources can be built and how they can fetch data from the internet through calls to Retrofit. Then, we will look at implementing local data sources and how they can interact with Room and Data Store to persist data locally. In the chapter's exercises, we will continue the previous exercises and add the data sources discussed in the chapter, seeing how we can connect them to Room and Retrofit.

In this chapter, we will cover the following topics:

- Building and using remote data sources
- Building and integrating local data sources

By the end of the chapter, you will have learned the role of data sources, how to implement remote and local data sources that use Retrofit, Room, and Data Store to manage an application's data, and how we can separate these data sources in separate library modules.

Technical requirements

The hardware and software requirements are as follows:

- Android Studio – Arctic Fox | 2020.3.1 Patch 3

The code files for this chapter can be found here: `https://github.com/ PacktPublishing/Clean-Android-Architecture/tree/main/Chapter7`.

Check out the following video to see the Code in Action: `https://bit.ly/3yOa7jE`

Building and using remote data sources

In this section, we will look at how we can build remote data sources and how we can use them in combination with Retrofit to fetch and manipulate data from the internet.

In the previous chapters, we defined abstractions for data sources that repositories depend on to manipulate data. This was because we wanted to avoid the repositories having dependencies on the data sources and instead have the data sources depend on the repositories. For remote data sources, this looks something like the following figure:

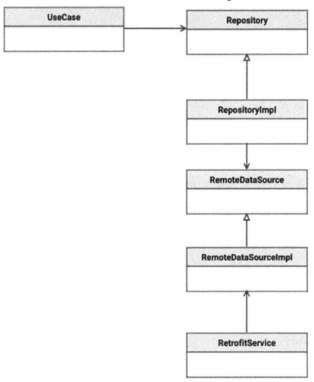

Figure 7.1 – A remote data source class diagram

The implementation of the remote data source has two roles. It will invoke the networking layer to fetch and manipulate data, and it will convert the data to either the domain entity or, if necessary, intermediary data required by the repository.

Let's look at the entity defined in the previous chapters:

```
data class User(
    val id: String,
    val firstName: String,
    val lastName: String,
    val email: String
) {

    fun getFullName() = "$firstName $lastName"
}
```

Here, we have the same `User` data class that was defined as part of the domain. Now let's assume we are fetching the following data from the internet in JSON format:

```
data class UserApiModel(
    @Json(name = "id") val id: String,
    @Json(name = "first_name") val firstName: String,
    @Json(name = "last_name") val lastName: String,
    @Json(name = "email") val email: String
)
```

Here, we have a `UserApiModel` class in which we define the same fields as the `User` class and annotate them with the `@Json` annotation, which is part of the Moshi library.

The remote data source abstraction looks like the following:

```
interface UserRemoteDataSource {

    fun getUser(id: String): Flow<User>
}
```

This is the abstraction we defined in the previous chapter. Before we write the implementation of this class, we will first need to specify our Retrofit service:

```
interface UserService {

    @GET("/users/{userId}")
```

```
    suspend fun getUser(@Path("userId") userId: String):
        UserApiModel
}
```

This is a typical Retrofit service class, which will fetch an `UserApiModel` class from the `/users/{userId}` endpoint. We can now create the implementation of the data source to fetch the user from `UserService`:

```
data class UserRemoteDataSourceImpl(private val userService:
UserService) : UserRemoteDataSource {
    override fun getUser(id: String): Flow<User> {
        return flow {
            emit(userService.getUser(id))
        }.map {
            User(it.id, it.firstName, it.lastName,
                it.email)
        }
    }
}
```

Here, we implement the `UserRemoteDataSource` interface, and in the `getUser` method, we invoke the `getUser` method from the `UserService` dependency. Once `UserApiModel` is obtained, we then convert it to the `User` class.

In this section, we looked at how we can build a remote data source with the help of the Retrofit library to manipulate data from the internet. In the section that follows, we will look at an exercise that shows how we can implement a remote data source.

Exercise 07.01 – Building a remote data source

Modify *Exercise 06.01 – Creating repositories* so that a new library module is created in Android Studio. Name the module `data-remote`. This module will depend on `domain` and `data-repository`. The module will be responsible for fetching users and posts as JSON from `https://jsonplaceholder.typicode.com/`.

The user will have the following JSON representation:

```
{
    "id": 1,
    "name": "Leanne Graham",
    "username": "Bret",
```

```
        "email": "Sincere@april.biz"
}
```

The post will have the following JSON representation:

```
{
    "userId": 1,
    "id": 1,
    "title": "sunt aut facere repellat provident
        occaecati excepturi optio reprehenderit",
    "body": "quia et suscipit\nsuscipit recusandae consequuntur
 expedita et cum\nreprehenderit molestiae ut ut quas totam\
 nnostrum rerum est autem sunt rem eveniet architecto"
}
```

The module will need to implement the following:

- `UserApiModel` and `PostApiModel`, which will hold the data from the JSON.
- `UserService`, which will return a list of `UserApiModel` from the /users URL and `UserApiModel` based on the ID from the /users/{userId} URL.
- `PostService`, which will return a list of `PostApiModel` from the /posts URL and `PostApiModel` based on the ID from the /post/{postId} URL.
- `RemoteUserDataSourceImpl`, which will implement `RemoteUserDataSource`, call `UserService`, and return `Flow`, which emits a list of `User` objects or `UseCaseException.UserException` if there is an error in the call to `UserService`. The same approach will be taken for returning `User` based on the ID.
- `RemotePostDataSourceImpl` which will implement `RemotePostDataSource`, call `PostService`, and return `Flow`, which emits a list of `Post` objects or `UseCaseException.PostException` if there is an error in the call to `PostService`. The same approach will be taken for returning a post based on the ID.

To complete this exercise, you will need to do the following:

1. Create the `data-remote` module.
2. Create the `UserApiModel` and `UserService` classes.
3. Create the `PostApiModel` and `PostService` classes.
4. Create the remote data sources implementations for `RemoteUserDataSource` and `RemotePostDataSource`.

Follow these steps to complete the exercise:

1. Create a new module named `data-remote`, which will be an Android library module.

2. Make sure that in the top-level `build.gradle` file, the following dependencies are set:

```
buildscript {

    ...

    dependencies {
        classpath gradlePlugins.android
        classpath gradlePlugins.kotlin
        classpath gradlePlugins.hilt
    }
}
```

3. In the same file, add the networking libraries to the library mappings:

```
    ext {

        ...

        versions = [

            ...

            okHttp              : "4.9.0",
            retrofit            : "2.9.0",
            moshi               : "1.13.0",

            ...

        ]

        ...

        network = [
            okHttp          : "com.squareup.okhttp3:
                okhttp:${versions.okHttp}",
            retrofit        : "com.squareup.retrofit2
                :retrofit:${versions.retrofit}",
            retrofitMoshi: "com.squareup.retrofit2
                :converter-moshi:$
                    {versions.retrofit}",
            moshi           : "com.squareup.moshi:
                moshi:${versions.moshi}",
            moshiKotlin  : "com.squareup.moshi:
```

```
                    moshi-kotlin:${versions.moshi}"
        ]
        ...
    }
```

4. In the `build.gradle` file of the `data-remote` module, make sure that the following plugins are present:

```
plugins {
    id 'com.android.library'
    id 'kotlin-android'
    id 'kotlin-kapt'
    id 'dagger.hilt.android.plugin'
}
```

5. In the same file, change the configurations to the ones defined in the top-level `build.gradle` file:

```
android {
    compileSdk defaultCompileSdkVersion

    defaultConfig {
        minSdk defaultMinSdkVersion
        targetSdk defaultTargetSdkVersion
        ...
    }
    compileOptions {
        sourceCompatibility javaCompileVersion
        targetCompatibility javaCompileVersion
    }
    kotlinOptions {
        jvmTarget = jvmTarget
    }
}
```

Here, we are making sure that the new module will use the same configurations with regards to compilation and the minimum and maximum Android version as the rest of the project, making it easier to change the configuration across all the modules.

6. In the same file, add the dependencies to the networking libraries and the `data-repository` and `domain` modules:

```
dependencies {
    implementation(project(path: ":domain"))
    implementation(project(path: ":data-repository"))
    implementation coroutines.coroutinesAndroid
    implementation network.okHttp
    implementation network.retrofit
    implementation network.retrofitMoshi
    implementation network.moshi
    implementation network.moshiKotlin
    implementation di.hiltAndroid
    kapt di.hiltCompiler
    testImplementation test.junit
    testImplementation test.coroutines
    testImplementation test.mockito
}
```

7. In the top-level `gradle.properties`, add the following configuration for `moshi`:

```
android.jetifier.ignorelist=moshi-1.13.0
```

8. In the `AndroidManifest.xml` file in the `data-remote` module, add the internet permission:

```
<?xml version="1.0" encoding="utf-8"?>
<manifest xmlns:android="http://schemas.android.com/apk/
res/android"
    package="com.clean.data_remote">
    <uses-permission android:name="android.permission.
INTERNET" />
</manifest>
```

9. In the `data-remote` module, create a new package called `networking`.

10. In the `networking` package, create a new package called `user`.

11. In the `user` package, create a new class called `UserApiModel`:

```
data class UserApiModel(
    @Json(name = "id") val id: Long,
```

```kotlin
    @Json(name = "name") val name: String,
    @Json(name = "username") val username: String,
    @Json(name = "email") val email: String
)
```

12. In the same package, create a new interface called `UserService`:

```kotlin
interface UserService {

    @GET("/users")
    suspend fun getUsers(): List<UserApiModel>

    @GET("/users/{userId}")
    suspend fun getUser(@Path("userId") userId: Long):
        UserApiModel
}
```

13. In the `networking` package, create a new package called `post`.

14. In the `post` package, create a new class called `PostApiModel`:

```kotlin
data class PostApiModel(
    @Json(name = "id") val id: Long,
    @Json(name = "userId") val userId: Long,
    @Json(name = "title") val title: String,
    @Json(name = "body") val body: String
)
```

15. In the same package, create a new interface called `PostService`:

```kotlin
interface PostService {

    @GET("/posts")
    suspend fun getPosts(): List<PostApiModel>

    @GET("/posts/{postId}")
    suspend fun getPost(@Path("postId") id: Long):
        PostApiModel
}
```

16. In the data-remote module, create a new package called source.

17. In the source package, create a new class called RemoteUserDataSourceImpl:

```
class RemoteUserDataSourceImpl @Inject
constructor(private val userService: UserService) :
    RemoteUserDataSource {

    override fun getUsers(): Flow<List<User>> = flow {
        emit(userService.getUsers())
    }.map { users ->
        users.map { userApiModel ->
            convert(userApiModel)
        }
    }.catch {
        throw UseCaseException.UserException(it)
    }

    override fun getUser(id: Long): Flow<User> = flow {
        emit(userService.getUser(id))
    }.map {
        convert(it)
    }.catch {
        throw UseCaseException.UserException(it)
    }

    private fun convert(userApiModel: UserApiModel) =
        User(userApiModel.id, userApiModel.name,
            userApiModel.username, userApiModel.email)
}
```

Here, we invoke the getUsers and getUser methods from UserService and then convert the UserApiModel objects to User objects to avoid the other layers depending on the networking-related data. The same principle applies to error handling. If there is a network error, such as an HTTP 404 code, the exception will be HttpException, which is part of the Retrofit library.

18. In the source package, create a new class called RemotePostDataSourceImpl:

```
class RemotePostDataSourceImpl @Inject
```

```
constructor(private val postService: PostService) :
    RemotePostDataSource {

    override fun getPosts(): Flow<List<Post>> = flow {
        emit(postService.getPosts())
    }.map { posts ->
        posts.map { postApiModel ->
            convert(postApiModel)
        }
    }.catch {
        throw UseCaseException.PostException(it)
    }

    override fun getPost(id: Long): Flow<Post> = flow {
        emit(postService.getPost(id))
    }.map {
        convert(it)
    }.catch {
        throw UseCaseException.PostException(it)
    }

    private fun convert(postApiModel: PostApiModel) =
        Post(postApiModel.id, postApiModel.userId,
            postApiModel.title, postApiModel.body)
}
```

Here, we follow the same principle as with the `RemoteUserDataSourceImpl` class.

19. In the `data-remote` module, create a new package called `injection`:

20. In the `injection` package, create a new class called `NetworkModule`:

```
@Module
@InstallIn(SingletonComponent::class)
class NetworkModule {

    @Provides
    fun provideOkHttpClient(): OkHttpClient =
```

```
            OkHttpClient
            .Builder()
            .readTimeout(15, TimeUnit.SECONDS)
            .connectTimeout(15, TimeUnit.SECONDS)
            .build()


    @Provides
    fun provideMoshi(): Moshi = Moshi.Builder().add
        (KotlinJsonAdapterFactory()).build()


    @Provides
    fun provideRetrofit(okHttpClient: OkHttpClient,
        moshi: Moshi): Retrofit = Retrofit.Builder()
        .baseUrl
            ("https://jsonplaceholder.typicode.com/")
        .client(okHttpClient)
        .addConverterFactory
            (MoshiConverterFactory.create(moshi))
        .build()


    @Provides
    fun provideUserService(retrofit: Retrofit):
        UserService =
        retrofit.create(UserService::class.java)


    @Provides
    fun providePostService(retrofit: Retrofit):
        PostService =
        retrofit.create(PostService::class.java)
}
```

Here, we provide the Retrofit and `OkHttp` dependencies required for networking.

21. In the `injection` package, create a class named `RemoteDataSourceModule`:

```
@Module
@InstallIn(SingletonComponent::class)
abstract class RemoteDataSourceModule {
```

```
    @Binds
    abstract fun bindPostDataSource(postDataSourceImpl:
  RemotePostDataSourceImpl): RemotePostDataSource

    @Binds
    abstract fun bindUserDataSource
        (userDataSourceImpl:
            RemoteUserDataSourceImpl):
  RemoteUserDataSource
}
```

Here, we use Hilt to bind the implementations from this module with the abstractions defined in the data-repository module.

22. To unit-test the code, we now need to create a new folder called resources in the test folder of the data-remote module.

23. Inside the resources folder, create a folder called mockito-extensions; inside this folder, create a file named org.mockito.plugins.MockMaker; and inside this file, add the following text – mock-maker-inline.

24. Create a test class named RemoteUserDataSourceImplTest, which will test the success scenarios for the methods inside RemoteUserDataSourceImpl:

```
class RemoteUserDataSourceImplTest {

    private val userService = mock<UserService>()
    private val userDataSource =
        RemoteUserDataSourceImpl(userService)

    @ExperimentalCoroutinesApi
    @Test
    fun testGetUsers() = runBlockingTest {
        val remoteUsers = listOf(UserApiModel(1,
            "name", "username", "email"))
        val expectedUsers = listOf(User(1, "name",
            "username", "email"))
        whenever(userService.getUsers()).
            thenReturn(remoteUsers)
```

```
        val result = userDataSource.getUsers().first()
        Assert.assertEquals(expectedUsers, result)
    }


    @ExperimentalCoroutinesApi
    @Test
    fun testGetUser() = runBlockingTest {
        val id = 1L
        val remoteUser = UserApiModel(id, "name",
            "username", "email")
        val user = User(id, "name", "username",
            "email")
        whenever(userService.getUser(id))
            .thenReturn(remoteUser)
        val result = userDataSource.getUser(id).
            first()
        Assert.assertEquals(user, result)
    }
}
```

Here, we are mocking the UserService interface and providing mock user data, which will then be obtained and converted by RemoteDataSourceImpl.

25. In the same test class, add the error scenarios:

```
class RemoteUserDataSourceImplTest {
    ...
    @ExperimentalCoroutinesApi
    @Test
    fun testGetUsersThrowsError() = runBlockingTest {
        whenever(userService.getUsers()).thenThrow
            (RuntimeException())
        userDataSource.getUsers().catch {
            Assert.assertTrue(it is UseCaseException.
                UserException)
        }.collect()
    }
```

```
    @ExperimentalCoroutinesApi
    @Test
    fun testGetUserThrowsError() = runBlockingTest {
        val id = 1L
        whenever(userService.getUser(id)).thenThrow
            (RuntimeException())
        userDataSource.getUser(id).catch {
            Assert.assertTrue(it is UseCaseException.
                UserException)
        }.collect()
    }
}
```

Here, we are mocking an error that is thrown by `UserService`, which will then be converted by `RemoteUserDataSourceImpl` into `UseCaseException.UserException`.

26. Create a test class named `RemotePostDataSourceImplTest`, which will have similar test methods as `RemoteUserDataSourceImplTest` for posts:

```
class RemotePostDataSourceImplTest {

    private val postService = mock<PostService>()
    private val postDataSource =
        RemotePostDataSourceImpl(postService)

    @ExperimentalCoroutinesApi
    @Test
    fun testGetPosts() = runBlockingTest {
        val remotePosts = listOf(PostApiModel(1, 1,
            "title", "body"))
        val expectedPosts = listOf(Post(1, 1, "title",
            "body"))
        whenever(postService.getPosts()).thenReturn
            (remotePosts)
        val result = postDataSource.getPosts().first()
        Assert.assertEquals(expectedPosts, result)
    }
```

```
@ExperimentalCoroutinesApi
@Test
fun testGetPost() = runBlockingTest {
    val id = 1L
    val remotePost = PostApiModel(id, 1, "title",
        "body")
    val expectedPost = Post(id, 1, "title",
        "body")
    whenever(postService.getPost(id)).thenReturn
        (remotePost)
    val result = postDataSource.getPost(id).
        first()
    Assert.assertEquals(expectedPost, result)
}
}
```

Here, we are doing for posts what we did for users in
`RemoteUserDataSourceImplTest`.

27. Add the error scenarios in `RemotePostDataSourceImplTest`:

```
class RemotePostDataSourceImplTest {
    …
    @ExperimentalCoroutinesApi
    @Test
    fun testGetPostsThrowsError() = runBlockingTest {
        whenever(postService.getPosts()).thenThrow
            (RuntimeException())
        postDataSource.getPosts().catch {
            Assert.assertTrue(it is UseCaseException.
                PostException)
        }.collect()
    }

    @ExperimentalCoroutinesApi
    @Test
    fun testGetPostThrowsError() = runBlockingTest {
```

```
        val id = 1L
        whenever(postService.getPost(id)).thenThrow
            (RuntimeException())
        postDataSource.getPost(id).catch {
            Assert.assertTrue(it is UseCaseException.
                PostException)
        }.collect()
    }
}
```

If we run the tests, we should see something like the following figure:

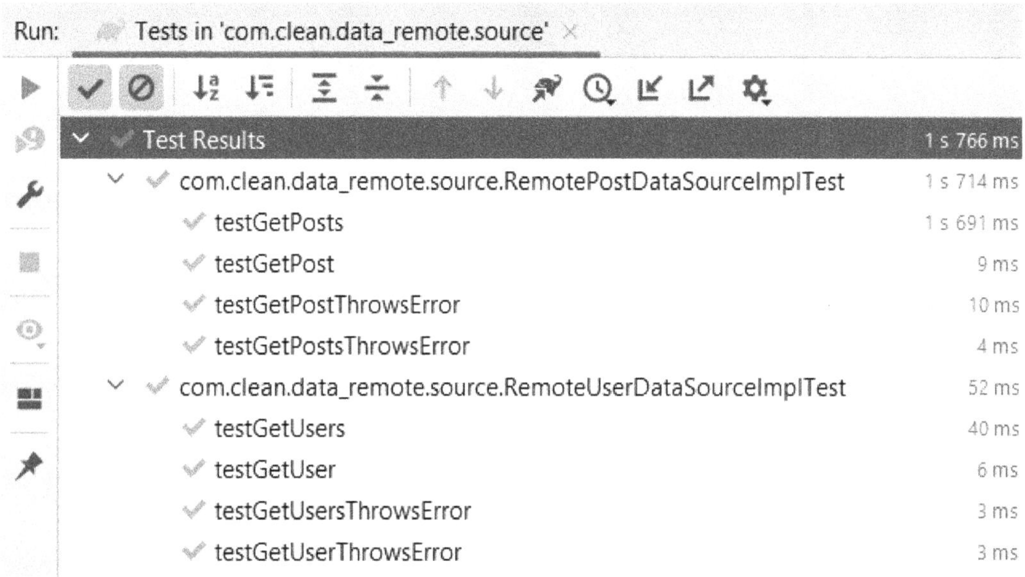

Figure 7.2 – Output of the remote data source unit tests

In this exercise, we have added a new module to the application, in which we can see how we can add a remote data source to the application. To fetch the data, we are using libraries such as OkHttp and Retrofit and combining them with the data source implementation for fetch users and posts. In the following section, we will expand the application to introduce local data sources, in which we will persist the data we are fetching here.

Building and integrating local data sources

In this section, we will analyze how we can build local data sources and integrate them with libraries such as Room and Data Store.

Local data sources have a similar structure to remote data sources. The abstractions are provided by the layers sitting above, and the implementations are responsible for invoking methods from persistence frameworks and converting data into entities, like the following figure:

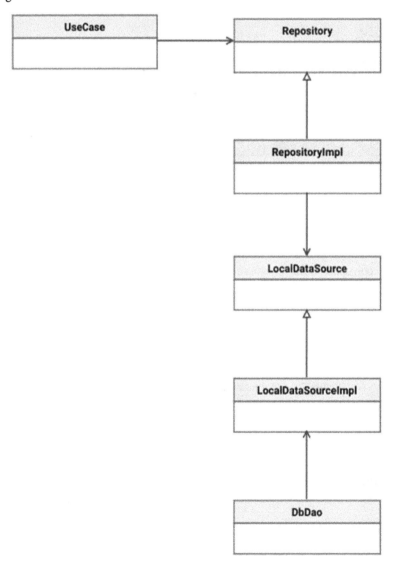

Figure 7.3 – A local data source diagram

Let's assume we have the same `UserEntity` defined in the previous chapters:

```
data class User(
    val id: String,
    val firstName: String,
    val lastName: String,
    val email: String
) {

    fun getFullName() = "$firstName $lastName"

}
```

Let's make the same assumption about `UserLocalDataSource`:

```
interface UserLocalDataSource {

    suspend fun insertUser(user: User)

    fun getUser(id: String): Flow<User>
}
```

We now need to provide an implementation for this data source that will manipulate the data from Room. First, we need to define a user entity for Room:

```
@Entity(tableName = "user")
data class UserEntity(
    @PrimaryKey @ColumnInfo(name = "id") val id: String,
    @ColumnInfo(name = "first_name") val firstName: String,
    @ColumnInfo(name = "last_name") val lastName: String,
    @ColumnInfo(name = "email") val email: String
)
```

Now, we can define `UserDao`, which queries a user by an ID and inserts a user:

```
@Dao
interface UserDao {

    @Query("SELECT * FROM user where id = :id")
    fun getUser(id: String): Flow<UserEntity>
```

```
    @Insert(onConflict = OnConflictStrategy.REPLACE)
    fun insertUser(users: UserEntity)
}
```

Finally, the implementation of the data source looks like this:

```
class UserLocalDataSourceImpl(private val userDao: UserDao) :
UserLocalDataSource {
    override suspend fun insertUser(user: User) {
        userDao.insertUser(UserEntity(user.id,
            user.firstName, user.lastName, user.email))
    }

    override fun getUser(id: String): Flow<User> {
        return userDao.getUser(id).map {
            User(it.id, it.firstName, it.lastName,
                it.email)
        }
    }
}
```

Here, the local data source invokes UserDao to insert and retrieve a user and converts the domain entity into a Room entity.

If we want to use Data Store instead of Room with a local data store implementation, we can have something like the following example:

```
private val KEY_ID = stringPreferencesKey("key_id")
private val KEY_FIRST_NAME =
    stringPreferencesKey("key_first_name")
private val KEY_LAST_NAME =
    stringPreferencesKey("key_last_name")
private val KEY_EMAIL = stringPreferencesKey("key_email")
class UserLocalDataSourceImpl(private val dataStore:
    DataStore<Preferences>) : UserLocalDataSource {
    override suspend fun insertUser(user: User) {
        dataStore.edit {
            it[KEY_ID] = user.id
```

```
            it[KEY_FIRST_NAME] = user.firstName
            it[KEY_LAST_NAME] = user.lastName
            it[KEY_EMAIL] = user.email
        }
    }


    override fun getUser(id: String): Flow<User> {
        return dataStore.data.map {
            User(
                it[KEY_ID].orEmpty(),
                it[KEY_FIRST_NAME].orEmpty(),
                it[KEY_LAST_NAME].orEmpty(),
                it[KEY_EMAIL].orEmpty()
            )
        }
    }
}
```

Here, we use a key for each of the fields of the User object to store the data. The getUser method doesn't use the ID to search for a user, which shows that for this particular use case, Room is the more appropriate method.

In this section, we looked at how we can build a local data source with the help of the Room and Data Store libraries to be able to query and persist data locally on a device. Next, we will look at an exercise to show how we can implement a local data store.

Exercise 07.02 – Building a local data source

Modify *Exercise 07.01 – Building a remote data source* so that a new Android library module named data-local is created. This module will depend on domain and data-repository.

The module will implement the following:

- UserEntity and PostEntity, which will hold data to be persisted from User and Post using Room

- UserDao and PostDao, which will be responsible for persisting and fetching a list of UserEntity and PostEntity

- `LocalUserDataSourceImpl` and `LocalPostDataSourceImpl`, which will be responsible for invoking the `UserDao` and `PostDao` objects to persist data and for converting data to `User` and `Post` objects
- `LocalInteractionDataSourceImpl`, which will be responsible for persisting the `Interaction` object

To complete this exercise, you will need to do the following:

1. Create the `data-local` module.
2. Create the `UserEntity` and `PostEntity` classes.
3. Create the DAOs for users and posts.
4. Create the data source implementations.

Follow these steps to complete the exercise:

1. Create a new module named `data-local`, which will be an Android library module.
2. Make sure that in the top-level `build.gradle` file, the following dependencies are set:

```
buildscript {

    ...

    dependencies {
        classpath gradlePlugins.android
        classpath gradlePlugins.kotlin
        classpath gradlePlugins.hilt
    }

}
```

3. In the same file, add the persistence libraries to the library mappings:

```
ext {

    ...

    versions = [

        ...

        room              : "2.4.0",
        datastore         : "1.0.0",

        ...

    ]
```

```
...
        persistence = [
                roomRuntime : "androidx.room:room-
                        runtime:${versions.room}",
                roomKtx      : "androidx.room:room-
                        ktx:${versions.room}",
                roomCompiler: "androidx.room:room-
                        compiler:${versions.room}",
                datastore    : "androidx.datastore:
                        datastore-preferences:$
                                {versions.datastore}"
        ]
        ...
}
```

4. In the `build.gradle` file of the `data-local` module, make sure that the following plugins are present:

```
plugins {
    id 'com.android.library'
    id 'kotlin-android'
    id 'kotlin-kapt'
    id 'dagger.hilt.android.plugin'
}
```

5. In the same file, change the configurations to the ones defined in the top-level `build.gradle` file:

```
android {
    compileSdk defaultCompileSdkVersion

    defaultConfig {
        minSdk defaultMinSdkVersion
        targetSdk defaultTargetSdkVersion

        ...
    }
    compileOptions {
        sourceCompatibility javaCompileVersion
```

```
            targetCompatibility javaCompileVersion
        }
        kotlinOptions {
            jvmTarget = jvmTarget
        }
    }
```

6. In the same file, add the dependencies to the networking libraries and the data-repository and domain modules:

```
dependencies {
    implementation(project(path: ":domain"))
    implementation(project(path: ":data-repository"))
    implementation coroutines.coroutinesAndroid
    implementation persistence.roomRuntime
    implementation persistence.roomKtx
    kapt persistence.roomCompiler
    implementation persistence.datastore
    implementation di.hiltAndroid
    kapt di.hiltCompiler
    testImplementation test.junit
    testImplementation test.coroutines
    testImplementation test.mockito
}
```

7. In the data-local module, create a new package called db.

8. In the db package, create a new package called user.

9. In the user package, create the UserEntity class:

```
@Entity(tableName = "user")
data class UserEntity(
    @PrimaryKey @ColumnInfo(name = "id") val id: Long,
    @ColumnInfo(name = "name") val name: String,
    @ColumnInfo(name = "username") val username:
        String,
    @ColumnInfo(name = "email") val email: String
)
```

10. In the same package, create the `UserDao` interface:

```
@Dao
interface UserDao {

    @Query("SELECT * FROM user")
    fun getUsers(): Flow<List<UserEntity>>

    @Insert(onConflict = OnConflictStrategy.REPLACE)
    fun insertUsers(users: List<UserEntity>)
}
```

11. In the db package, create a new package called `post`.

12. In the `post` package, create a new class called `PostEntity`:

```
@Entity(tableName = "post")
data class PostEntity(
    @PrimaryKey @ColumnInfo(name = "id") val id: Long,
    @ColumnInfo(name = "userId") val userId: Long,
    @ColumnInfo(name = "title") val title: String,
    @ColumnInfo(name = "body") val body: String
)
```

13. In the same package, create a new interface called `PostDao`:

```
@Dao
interface PostDao {

    @Query("SELECT * FROM post")
    fun getPosts(): Flow<List<PostEntity>>

    @Insert(onConflict = OnConflictStrategy.REPLACE)
    fun insertPosts(users: List<PostEntity>)
}
```

14. In the db package, create the `AppDatabase` class:

```
@Database(entities = [UserEntity::class,
PostEntity::class], version = 1)
abstract class AppDatabase : RoomDatabase() {
```

```
        abstract fun userDao(): UserDao

        abstract fun postDao(): PostDao
}
```

15. In the `data-local` module, create a new package called `source`.

16. In the `source` package, create a new class called `LocalUserDataSourceImpl`:

```
class LocalUserDataSourceImpl @Inject constructor(private
val userDao: UserDao) :
    LocalUserDataSource {

        override fun getUsers(): Flow<List<User>> =
            userDao.getUsers().map { users ->
            users.map {
                User(it.id, it.name, it.username,
                    it.email)
            }
        }

        override suspend fun addUsers(users: List<User>) =
            userDao.insertUsers(users.map {
            UserEntity(it.id, it.name, it.username,
                it.email)
        })
}
```

Here, in the `getUsers` method, we retrieve a list of `UserEntity` objects from `UserDao` and convert them into `User` objects. In the `addUsers` method, we do the opposite, by taking a list of `User` objects to be inserted and converting them into `UserEntity` objects.

17. In the same package, create the `LocalPostDataSourceImpl` class:

```
class LocalPostDataSourceImpl @Inject constructor(private
val postDao: PostDao) :
    LocalPostDataSource {
        override fun getPosts(): Flow<List<Post>> =
            postDao.getPosts().map { posts ->
```

```
        posts.map {
            Post(it.id, it.userId, it.title, it.body)
        }
    }

    override suspend fun addPosts(posts: List<Post>) =
        postDao.insertPosts(posts.map {
        PostEntity(it.id, it.userId, it.title,
            it.body)
    })
}
```

Here, we follow the same approach we used for `LocalUserDataSourceImpl`.

18. In the same package, create the `LocalInteractionDataSourceImpl` class:

```
internal val KEY_TOTAL_TAPS = intPreferencesKey("key_
total_taps")
class LocalInteractionDataSourceImpl @
Inject constructor(private val dataStore:
DataStore<Preferences>) :
    LocalInteractionDataSource {

    override fun getInteraction(): Flow<Interaction> {
        return dataStore.data.map {
            Interaction(it[KEY_TOTAL_TAPS] ?: 0)
        }
    }

    override suspend fun saveInteraction(interaction:
        Interaction) {
        dataStore.edit {
            it[KEY_TOTAL_TAPS] =
                interaction.totalClicks
        }
    }
}
```

Here, we use the Preference Data Store library to persist the Interaction object, by holding different keys for each field in the `Interaction` class, and in this case, it will be just one key for the total clicks.

19. In the `data-local` module, create a new package named `injection`.

20. In the `injection` package, create a new class named `PersistenceModule`:

```
val Context.dataStore: DataStore<Preferences> by
preferencesDataStore(name = "my_preferences")
@Module
@InstallIn(SingletonComponent::class)
class PersistenceModule {

    @Provides
    fun provideAppDatabase(@ApplicationContext
        context: Context): AppDatabase =
    Room.databaseBuilder(
        context,
        AppDatabase::class.java, "my-database"
    ).build()

    @Provides
    fun provideUserDao(appDatabase: AppDatabase):
        UserDao = appDatabase.userDao()

    @Provides
    fun providePostDao(appDatabase: AppDatabase):
        PostDao = appDatabase.postDao()

    @Provides
    fun provideLocalInteractionDataSourceImpl
        (@ApplicationContext context: Context) =
        LocalInteractionDataSourceImpl(context.dataStore)
}
```

Here, we provide all the Data Store and Room dependencies.

21. In the same package, create a new class called `LocalDataSourceModule`, in which we connect the abstractions to the bindings:

```
@Module
@InstallIn(SingletonComponent::class)
abstract class LocalDataSourceModule {

    @Binds
    abstract fun bindPostDataSource
        (lostDataSourceImpl: LocalPostDataSourceImpl):
            LocalPostDataSource

    @Binds
    abstract fun bindUserDataSource
        (userDataSourceImpl: LocalUserDataSourceImpl):
            LocalUserDataSource

    @Binds
    abstract fun bindInteractionDataStore
        (interactionDataStore:LocalInteractionData
            SourceImpl): LocalInteractionDataSource
}
```

22. To unit-test the code, we will now need to create a new folder called `resources` in the test folder of the `data-local` module.

23. Inside the `resources` folder, create a folder called `mockito-extensions`; inside this folder, create a file named `org.mockito.plugins.MockMaker`; and inside this file, add the following text – `mock-maker-inline`.

24. Create the `LocalUserDataSourceImplTest` test class:

```
class LocalUserDataSourceImplTest {

    private val userDao = mock<UserDao>()
    private val userDataSource =
        LocalUserDataSourceImpl(userDao)

    @ExperimentalCoroutinesApi
    @Test
```

```
fun testGetUsers() = runBlockingTest {
    val localUsers = listOf(UserEntity(1, "name",
        "username", "email"))
    val expectedUsers = listOf(User(1, "name",
        "username", "email"))
    whenever(userDao.getUsers()).thenReturn
        (flowOf(localUsers))
    val result = userDataSource.getUsers().first()
    Assert.assertEquals(expectedUsers, result)
}


@ExperimentalCoroutinesApi
@Test
fun testAddUsers() = runBlockingTest {
    val localUsers = listOf(UserEntity(1, "name",
        "username", "email"))
    val users = listOf(User(1, "name", "username",
        "email"))
    userDataSource.addUsers(users)
    verify(userDao).insertUsers(localUsers)
}
}
```

Here, we are mocking the UserDao class and using it to provide mock data to
LocalUserDataSourceImpl, which will then convert the data to and from the
User objects.

25. Create the LocalPostDataSourceImplTest test class:

```
class LocalPostDataSourceImplTest {

    private val postDao = mock<PostDao>()
    private val postDataSource =
        LocalPostDataSourceImpl(postDao)


    @ExperimentalCoroutinesApi
    @Test
    fun testGetPosts() = runBlockingTest {
```

```
        val localPosts = listOf(PostEntity(1, 1,
            "title", "body"))
        val expectedPosts = listOf(Post(1, 1, "title",
            "body"))
        whenever(postDao.getPosts()).thenReturn
            (flowOf(localPosts))
        val result = postDataSource.getPosts().first()
        Assert.assertEquals(expectedPosts, result)
    }


    @ExperimentalCoroutinesApi
    @Test
    fun testAddUsers() = runBlockingTest {
        val localPosts = listOf(PostEntity(1, 1,
            "title", "body"))
        val posts = listOf(Post(1, 1, "title",
            "body"))
        postDataSource.addPosts(posts)
        verify(postDao).insertPosts(localPosts)
    }
}
```

Here, we perform the same type of tests for posts as we did in
LocalUserDataSourceImplTest for users.

26. Create the LocalInteractionDataSourceImplTest test class:

```
class LocalInteractionDataSourceImplTest {

    private val dataStore = mock<DataStore
        <Preferences>>()
    private val interactionDataSource =
        LocalInteractionDataSourceImpl(dataStore)

    @ExperimentalCoroutinesApi
    @Test
    fun testGetInteraction() = runBlockingTest {
        val clicks = 10
```

```
        val interaction = Interaction(clicks)
        val preferences = mock<Preferences>()
        whenever(preferences[KEY_TOTAL_TAPS]).
            thenReturn(clicks)
        whenever(dataStore.data).thenReturn
            (flowOf(preferences))
        val result = interactionDataSource.
            getInteraction().first()
        assertEquals(interaction, result)
    }


    @ExperimentalCoroutinesApi
    @Test
    fun testSaveInteraction() = runBlockingTest {
        val clicks = 10
        val interaction = Interaction(clicks)
        val preferences = mock<MutablePreferences>()
        whenever(preferences.toMutablePreferences())
            .thenReturn(preferences)
        whenever(dataStore.updateData(any())).
            thenAnswer {
            runBlocking {
                it.getArgument<suspend (Preferences) -
                > Preferences>(0).invoke(preferences)
            }
            preferences
        }
        interactionDataSource.saveInteraction(interaction)
        verify(preferences)[KEY_TOTAL_TAPS] = clicks
    }
}
```

Here, in the testSaveInteraction method, we need to mock the
updateData method instead of the edit method from the DataStore class.
This is because the edit method is an extension function that can't be mocked
with the current libraries we have and instead must rely on the method it invokes,
which is updateData.

If we run the tests, we should see something like the following figure:

Run:	Tests in 'com.clean.data_local.source' ×	

	Test Results	1 s 965 ms
	com.clean.data_local.source.LocalInteractionDataSourceImplTest	1 s 878 ms
	testSaveInteraction	1 s 835 ms
	testGetInteraction	43 ms
	com.clean.data_local.source.LocalPostDataSourceImplTest	37 ms
	testGetPosts	35 ms
	testAddUsers	2 ms
	com.clean.data_local.source.LocalUserDataSourceImplTest	50 ms
	testGetUsers	48 ms
	testAddUsers	2 ms

Figure 7.4 – Output of the local data source unit tests

If we draw a diagram of the modules in the exercise, we will see something like the following figure:

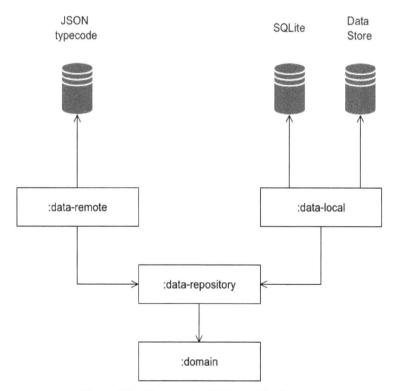

Figure 7.5 – The exercise 07.02 module diagram

We can see that the `:data-remote` and `:data-local` modules are isolated from each other. The two modules have different responsibilities and deal with different dependencies. `:data-remote` deals with fetching data from the internet, while `:data-local` deals with persisting data locally into SQLite using Room and files using Data Store. This gives our code more flexibility because we are able to change how we fetch data – for example, without impacting how we persist the data.

In this exercise, we have created a new module in the application in which we deal with local data sources. To persist data, we have used libraries such as Room and Data Store, and we have integrated them with the local data store.

Summary

In this chapter, we looked at the concept of data sources and the different types of data sources we have available in an Android application. We started with remote data sources and saw some examples of how we can build a data source and combine it with libraries such as Retrofit and OkHttp. The local data source followed similar principles as the remote one, and here, we have used libraries such as Room and Data Store to implement this.

In the exercises, we implemented the data sources as part of different modules. This was to avoid creating any unnecessary dependencies between the other layers of the application and the specific frameworks we have used for the data sources. In the next chapter, we will look at how we can build the presentation layer and show data to the user. We will also explore how we can split the presentation layer into separate modules and navigate from a screen in one module to a screen in another module, through the introduction of modules that can be shared by other presentation modules.

Part 3 –
Presentation Layer

This part will go over the presentation layer and the patterns that can be applied to have a decoupled and testable code base.

This part includes the following chapters:

8

Implementing an MVVM Architecture

In this chapter, we will look at how data can be presented by Android applications to end users. We will look over the available architecture patterns for data presentation and analyze the differences between them. Later, we will look at the **Model-View-ViewModel (MVVM)** pattern, the role it plays in separating business logic and user interface updates, and how we can implement it using **Android Architecture Components**. Finally, we will look at how we can split the presentation layer across multiple library modules. In the exercises of this chapter, we will integrate the layers built in the previous chapters with a presentation layer built using MVVM, we will create a presentation layer that will plug into the domain layer to fetch and update the data, and we will also look at how we handle common logic between different modules in the presentation layer.

In this chapter, we will cover the following topics:

- Presenting data in Android applications
- Presenting data with MVVM
- Presenting data in multiple modules

By the end of the chapter, you will be able to implement the MVVM architecture pattern in an Android application using the ViewModel architecture component and be able to split the presentation layer into separate library modules.

Technical requirements

This chapter has the following hardware and software requirements:

- Android Studio Arctic Fox 2020.3.1 Patch 3

The code files for this chapter can be found here: `https://github.com/PacktPublishing/Clean-Android-Architecture/tree/main/Chapter8`.

Check out the following video to see the Code in Action: `https://bit.ly/3FZJWI1`

Presenting data in Android applications

In this section, we will look at various architecture patterns suitable for presenting data in an Android application and analyze their benefits and drawbacks.

Early Android applications relied on a pattern similar to the **Model-View-Controller (MVC)** architecture pattern, where an activity is the Controller, the View is represented by the `android.widget.View` hierarchy, and the Model is responsible for managing the application's data. The relationship between the components would look something like the following:

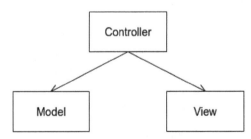

Figure 8.1 – Android MVC relationship

From *Figure 8.1*, we can see that the Controller represented by the activity would interact with the Model to fetch and manipulate the data, and then it would update the View with the relevant information.

The idea is to have each `Activity` sandboxed as much as possible so that they can be offered and shared between multiple applications (like how the Camera application is opened by other applications to take photos and offer those photos to those applications). Because of this, activities need to be started using intents and not by instantiating them. By removing the ability to instantiate an `Activity` directly, we lose the ability to inject dependencies through the constructor. Another factor we need to consider is that activities have life cycle states, and we inherit these states in each `Activity` in our application. All these factors combined make an `Activity` very hard or next to impossible to unit test unless we use a library such as **Robolectric** or rely on instrumented tests on an Android device or emulator. Both options are slow and, in the case of instrumented tests, can be expensive when we need to run the tests in testing clouds such as **Firebase Test Lab**.

To solve the problem of unit testing logic that was present in activities and later fragments, various adaptations of the **Humble Object** pattern emerged. More information about the pattern can be found here: `http://xunitpatterns.com/Humble%20Object.html`. The idea was to separate as much as possible the logic present in activities into separate objects and unit test those objects. One of the most popular solutions was the **Model-View-Presenter** (**MVP**) architecture pattern. In this pattern, the `Activity` along with the `android.widget.View` hierarchy becomes the View, the Presenter is responsible for fetching the data from the model and performing the logic required, updating the View, and the Model has the same responsibility as in MVC to handle the application's data. The relationship between these components looks like the following figure:

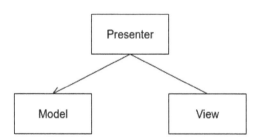

Figure 8.2 – MVP relationship

The interesting aspect of the relationship between the components is the relationship between the Presenter and the View, which goes both ways. The Presenter will update the View with the relevant data, but the View will also invoke the Presenter, if necessary for user interactions. Because of the relationship between the two components, the definition of a contract is required, which looks like the following:

```
interface Presenter {
```

```
    fun loadUsers()

    fun validateInput(text: String)
}

interface View {

    fun showUsers(users: List<User>)

    fun showInputError(error: String)
}
```

Here, we have a `View` interface and a `Presenter` interface. The implementation of the `Presenter` might look something like this:

```
class PresenterImpl(
    private val view: View,
    private val getUsersUseCase: GetUsersUseCase
) : Presenter {

    private val scope = CoroutineScope(Dispatchers.Main)

    override fun loadUsers() {
        scope.launch {
            getUsersUseCase.execute()
                .collect { users ->
                    view.showUsers(users)
                }
        }

    }

    override fun validateInput(text: String) {
        if (text.isEmpty()) {
            view.showInputError("Invalid input")
        }
    }
}
```

Here, the `PresenterImpl` class has a dependency on the `View` and on a `GetUsersUseCase` object, which will return a `Flow` object containing a list of users. When the `Presenter` receives the list of users, it will call the `showUsers` method from the `View`. When the `validateInput` method is called, the `Presenter` will check whether the text is empty and invoke the `showInputError` method from the `View` with an error message. The implementation of the `View` might look like the following:

```
class MainActivity : ComponentActivity(), View {
    @Inject
    private lateinit var presenter: Presenter
    private lateinit var usersAdapter: UsersAdapter
    private lateinit var editText: EditText
    private lateinit var errorView: TextView

    override fun onCreate(savedInstanceState: Bundle?) {
        super.onCreate(savedInstanceState)
        …
        editText.addTextChangedListener(object :
            TextWatcher {
                …
                override fun afterTextChanged(s: Editable?) {
                    presenter.validateInput(s?.toString().
orEmpty())
                }

        })
        presenter.loadUsers()

    }

    override fun showUsers(users: List<User>) {
        usersAdapter.add(users)
    }

    override fun showInputError(error: String) {
        errorView.text = error
    }
}
```

Here, we implement the `View` interface in `MainActivity`; in the implementation of the methods, we call the appropriate `View`- related classes to show the relevant data, such as showing the error message for an invalid input in a `TextView` object and setting the data in a `RecyclerView.Adapter` object. For validating the input, when the text changes in an `EditText` object, it will invoke the `Presenter` to validate the new text. The `Presenter` dependency will be injected using some form of dependency injection.

Because presenters will end up performing background operations, we run the risk of causing `Context` leaks. This means that we need to factor the life cycle of the `Activity` into the MVP contract. To achieve this, we will need to define a `close` method in the `Presenter`:

```
interface Presenter {
    …
    fun close()
}
```

In the preceding snippet, we added the `close` method, which will be called in the `onDestroy` method of the `Activity` as follows:

```
override fun onDestroy() {
        presenter.close()
        super.onDestroy()
}
```

The implementation of the `close` method will have to clean up all the resources that might cause any leaks:

```
class PresenterImpl(
    private val view: View,
    private val getUsersUseCase: GetUsersUseCase
) : Presenter {

    private val scope = CoroutineScope(Dispatchers.Main)
    …
    override fun close() {
        scope.cancel()
    }
}
```

Here, we are canceling the subscription to the `Flow` object so that we will not receive any updates after the `Activity` is destroyed.

In this section, we have looked at previous architecture patterns used in Android applications, from the MVC-like approach that was used in early Android applications to MVP, which aimed to solve some of the problems of the initial approach. Although MVP was popular in the past and is still present in some Android applications, it has slowly been phased out, mainly because of the release of Android Architecture Components, which rely on the MVVM pattern, and additionally, Jetpack Compose, which works better with data flows, which are more suited to MVVM. In the section that follows, we will look at the MVVM architecture pattern and how it is different from MVP as a concept.

Presenting data with MVVM

In this section, we will analyze the **Model-View-ViewModel** architecture pattern and how it is implemented for Android applications.

MVVM represents a different approach to the Humble Object pattern, which attempts to extract the logic out of activities and fragments. In MVVM, the View is represented by activities and fragments as it was in MVP, the Model plays the same role, managing the data, and the ViewModel sits between the two by requesting the data from the Model when the View requires it. The relationship between the three is as follows:

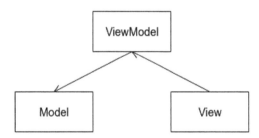

Figure 8.3 – MVVM relationship

In *Figure 8.3*, we see a unidirectional relationship between the three components. The View has a dependency on the ViewModel, and the ViewModel has a dependency on the Model. This allows for more flexibility because multiple Views can use the same ViewModel. For the data to be updated in the View, MVVM requires an implementation of the **Observer** pattern. This means that the ViewModel uses an **Observable**, which the View will subscribe to and react to changes in the data.

To develop Android applications, we have the possibility of using the Android Architecture Components libraries, which provide a `ViewModel` class that solves the issue of activity and fragment life cycles, combined with coroutine extensions useful for subscribing to flows or coroutines to stop the emission of data when the activities and fragments are in invalid states for data to be displayed and to avoid context leaks.

From the perspective of **Clean Architecture**, MVVM sits on the **Interface Adapter** layer. It has the role of fetching the data from the **Use Case** layer and converting the entities into objects that the **Framework** layer requires. It also handles changes to the data triggered by the user and converts this data back into entities, passing it back to the Use Case layer. In *Chapter 3, Understanding Data Presentation on Android*, we discussed the Android Architecture Components libraries and saw how we can implement ViewModels combined with `LiveData` (which acts as the observable that the View can subscribe to). An example of a `ViewModel` class might look like the following:

```kotlin
class MyViewModel(
    private val getUsersUseCase: GetUsersUserUseCase
) : ViewModel() {

    private val _usersFlow =
        MutableStateFlow<List<UiUser>>(listOf<UiUser>())
    val usersFlow: StateFlow<List<UiUser>> = _usersFlow

    fun load() {
        viewModelScope.launch {
            getUsersUseCase.execute()
                .map {
                    // Convert List<User> to List<UiUser>
                }
                .collect {
                    _usersFlow.value = it
                }
        }
    }
}
```

Here, we load a list of `User` objects and then keep that list inside a `StateFlow` object. This `StateFlow` object replaces `LiveData` and represents the observable that the View will subscribe to. When the View requires the list of users, it will invoke the `load` method.

In this section, we have analyzed the MVVM architecture pattern and the difference between it and the MVP pattern. In the following section, we will look at how we can present data using MVVM inside an Android application.

Exercise 08.01 – Implementing MVVM

Modify *Exercise 07.02*, *Building a local data source*, of *Chapter 7*, *Building Data Sources*, so that a new module called `presentation-posts` is created. The module will be responsible for displaying the data from `GetPostsWithUsersWithInteractionUseCase` using MVVM. The data will be displayed in the following format:

- A header with the following text: "Total click count: x" where x is the number of clicks taken from the `totalClicks` field in the `Interaction` class

- A list of posts where each row contains the following: "Author: x" and "Title: y" where x is the `name` field in the `User` class, and y is the `title` field in the `Post` class

- A loading view for when the data is being loaded

- A `Snackbar` view for when there is an error

To complete this exercise, you will need to do the following:

1. Create the `presentation-post` module.

2. Create a new sealed class called `UiState`, which will have as subclasses `Loading`, `Error` (which will hold an error message), and `Success` (which will hold the post data).

3. Create a new class called `PostListItemModel`, which will have `id`, `author`, and `name` as fields.

4. Create a new class called `PostListModel`, which will have a `headerText` field and a list of `PostListItemModel` objects.

5. Create a new class called `PostListConverter`, which will convert a `Result.Success` object into a `UiState.Success`, which holds the `PostListModel` object and will convert a `Result.Error` object into a `UiState.Error` object.

6. Create a new class called `PostListViewModel`, which will load the data from `GetPostsWithUsersWithInteractionUseCase`, convert the data using `PostListConverter`, and store `UiState` in `StateFlow`.

7. Create a new Kotlin file, which will contain `@Composable` methods responsible for drawing the UI.

8. Modify `MainActivity` in the app module so that it will display the list of posts.

Follow these steps to complete the exercise:

1. Create a new module called `presentation-post`, which will be an Android library module.

2. Make sure that in the top-level `build.gradle` file, the following dependencies are set:

```
buildscript {

    ...

    dependencies {
        classpath gradlePlugins.android
        classpath gradlePlugins.kotlin
        classpath gradlePlugins.hilt
    }
}
```

3. In the same file, add the persistence libraries to the library mappings:

```
buildscript {
    ext {

        ...

        versions = [

            ...

            viewModel              : "2.4.0",
            navigationCompose      : "2.4.0-rc01",
            hiltNavigationCompose: "1.0.0-rc01",

            ...

        ]

        ...

        androidx = [

            ...

            viewModelKtx              : "androidx.
                lifecycle:lifecycle-viewmodel-
                    ktx:${versions.viewModel}",
            viewModelCompose          : "androidx.
                lifecycle:lifecycle-viewmodel-
                    compose:${versions.viewModel}",
            navigationCompose         : "androidx.
```

```
            navigation:navigation-compose:$
               {versions.navigationCompose}",
        hiltNavigationCompose   : "androidx.
            hilt:hilt-navigation-compose:$
               {versions.hiltNavigationCompose}"
    ]
    ...
    }
  ...
  }
```

Here, we have added dependencies for the ViewModel library as well as the Navigation library (which will be used in later exercises).

4. In the `build.gradle` file of the `presentation-post` module, make sure that the following plugins are present:

```
plugins {
    id 'com.android.library'
    id 'kotlin-android'
    id 'kotlin-kapt'
    id 'dagger.hilt.android.plugin'
}
```

5. In the same file, change the configurations to the ones defined in the top-level `build.gradle` file:

```
android {
    compileSdk defaultCompileSdkVersion

    defaultConfig {
        minSdk defaultMinSdkVersion
        targetSdk defaultTargetSdkVersion
        ...
    }

    ...

    compileOptions {
        sourceCompatibility javaCompileVersion
```

```
        targetCompatibility javaCompileVersion
    }
    kotlinOptions {
        jvmTarget = jvmTarget
        useIR = true
    }
    buildFeatures {
        compose true
    }
    composeOptions {
        kotlinCompilerExtensionVersion versions.
            compose
    }
}
```

Here, we keep the same configuration consistent with the other modules in the application, and we integrate the Jetpack Compose configuration.

6. In the same file, add the dependencies to the networking libraries and domain modules:

```
dependencies {
    implementation(project(path: ":domain"))
    implementation coroutines.coroutinesAndroid
    implementation androidx.composeUi
    implementation androidx.composeMaterial
    implementation androidx.viewModelKtx
    implementation androidx.viewModelCompose
    implementation androidx.lifecycleRuntimeKtx
    implementation androidx.navigationCompose
    implementation di.hiltAndroid
    kapt di.hiltCompiler
    testImplementation test.junit
    testImplementation test.coroutines
    testImplementation test.mockito
}
```

7. In the presentation-post module, create a package called list.

8. In the `list` package, create the `UiState` class:

```
sealed class UiState<T : Any> {
    object Loading : UiState<Nothing>()
    data class Error<T : Any>(val errorMessage:
        String) : UiState<T>()
    data class Success<T : Any>(val data: T) :
        UiState<T>()
}
```

9. In the same package, create a file called `PostListModels`.

10. In the `PostListModels` file, create the `PostListItemModel` class:

```
data class PostListItemModel(
    val id: Long,
    val userId: Long,
    val authorName: String,
    val title: String
)
```

11. In the same file, create the `PostListModel` class:

```
data class PostListModel(
    val headerText: String = "",
    val items: List<PostListItemModel> = listOf()
)
```

12. In the `presentation-post` module, in the `src/main` folder, create a folder called `res`.

13. In the `res` folder, create a new folder called `values`.

14. In the `values` folder, create a file called `strings.xml`.

15. In the `strings.xml` file, add the following strings:

```
<?xml version="1.0" encoding="utf-8"?>
<resources>
    <string name="total_click_count">Total click
        count: %d</string>
    <string name="author">Author: %s</string>
    <string name="title">Title: %s</string>
</resources>
```

16. In the `list` package, create the `PostListConverter` class:

```
class PostListConverter @Inject constructor(@
ApplicationContext private val context: Context) {

    fun convert(postListResult: Result
        <GetPostsWithUsersWithInteractionUseCase.
            Response>): UiState<PostListModel> {
        return when (postListResult) {
            is Result.Error -> {
                UiState.Error(postListResult.
                    exception.localizedMessage.orEmpty())
            }
            is Result.Success -> {
                UiState.Success(PostListModel(
                    headerText = context.getString(
                        R.string.total_click_count,
                        postListResult.data.
                            interaction.totalClicks
                    ),
                    items = postListResult.data.
                    posts.map {
                    PostListItemModel(
                        it.post.id,
                        it.user.id,
                        context.getString(R.string.
author, it.user.name),
                        context.getString(R.string.
title, it.post.title)
                    )
                }
            ))
            }
        }
    }
}
```

Here, we convert the `Result.Success` and `Result.Error` objects into equivalent `UiState` objects, which will be used to display the information to the user.

17. In the `list` package, create the `PostListViewModel` class:

```
@HiltViewModel
class PostListViewModel @Inject constructor(
    private val useCase:
        GetPostsWithUsersWithInteractionUseCase,
    private val converter: PostListConverter
) : ViewModel() {

    private val _postListFlow =
        MutableStateFlow<UiState
            <PostListModel>>(UiState.Loading)
    val postListFlow:
        StateFlow<UiState<PostListModel>> =
            _postListFlow

    fun loadPosts() {
        viewModelScope.launch {
            useCase.execute
                (GetPostsWithUsersWithInteractionUseCase
                    .Request)
                .map {
                    converter.convert(it)
                }
                .collect {
                    _postListFlow.value = it
                }
        }
    }
}
```

Here, we get the list of posts and users from the `GetPostsWithUsersInteractionUseCase` object, then we convert it to the `UiState` object, and finally, we update `StateFlow` with the `UiState` object.

18. In the `list` package, create a file called `PostListScreen`.

19. In the `PostListScreen` file, add a method to display a loading widget and a `Snackbar` method:

```kotlin
@Composable
fun Error(errorMessage: String) {
    Column(
        modifier = Modifier.fillMaxSize(),
        verticalArrangement = Arrangement.Bottom
    ) {
        Snackbar {
            Text(text = errorMessage)
        }
    }
}

@Composable
fun Loading() {
    Column(
        modifier = Modifier.fillMaxSize(),
        verticalArrangement = Arrangement.Center,
        horizontalAlignment =
            Alignment.CenterHorizontally,
    ) {
        CircularProgressIndicator()
    }
}
```

20. In the same file, add a method to display the list of posts and the header:

```kotlin
@Composable
fun PostList(
    postListModel: PostListModel
) {
    LazyColumn(modifier = Modifier.padding(16.dp)) {
        item(postListModel.headerText) {
            Column(modifier = Modifier.padding(16.dp)) {
                Text(text = postListModel.headerText)
```

```
                }
            }
        items(postListModel.items) { item ->
            Column(
                modifier = Modifier
                    .padding(16.dp)
            ) {
                Text(text = item.authorName)
                Text(text = item.title)
            }
        }
    }
}
```

21. In the same file, add a method that will monitor the value of `postListFlow` and invoke one of the preceding three methods, depending on the value of the state:

```
@Composable
fun PostListScreen(
    viewModel: PostListViewModel
) {
    viewModel.loadPosts()
    viewModel.postListFlow.collectAsState().value.let {
state ->
        when (state) {
            is UiState.Loading -> {
                Loading()
            }
            is UiState.Error -> {
                Error(state.errorMessage)
            }
            is UiState.Success -> {
                PostList(state.data)
            }
        }
    }
}
```

22. In the `build.gradle` file of the `app` module, make sure that the following plugins are added:

```
plugins {
    id 'com.android.application'
    id 'kotlin-android'
    id 'kotlin-kapt'
    id 'dagger.hilt.android.plugin'
}
```

23. In the same file, make sure that the following dependencies are added:

```
dependencies {
    implementation(project(path: ":presentation-
        post"))
    implementation(project(path: ":domain"))
    implementation(project(path: ":data-remote"))
    implementation(project(path: ":data-local"))
    implementation(project(path: ":data-repository"))
    implementation androidx.core
    implementation androidx.appCompat
    implementation material.material
    implementation androidx.composeUi
    implementation androidx.composeMaterial
    implementation androidx.composeUiToolingPreview
    implementation androidx.lifecycleRuntimeKtx
    implementation androidx.composeActivity
    implementation androidx.navigationCompose
    implementation androidx.hiltNavigationCompose
    implementation di.hiltAndroid
    kapt di.hiltCompiler
    testImplementation test.junit
}
```

24. In the `app` module, create a package called `injection`.

25. In the `injection` package, create a class called `AppModule`:

```
@Module
@InstallIn(SingletonComponent::class)
```

```
class AppModule {

    @Provides
    fun provideUseCaseConfiguration() =
        UseCase.Configuration(Dispatchers.IO)
}
```

Here, we provide a `UseCase.Configuration` dependency, which will be injected into all the `UseCase` subclasses.

26. In the app module, create a class called `PostApplication`:

```
@HiltAndroidApp
class PostApplication : Application()
```

27. Add the `PostApplication` class to the `AndroidManifest.xml` file of the app module:

```
<application

    ...

    android:name=".PostApplication"

    ...

    >
```

28. Modify the `MainActivity` class so that it will use the navigation library to go to the `PostListScreen` function from the `presentation-post` module:

```
@AndroidEntryPoint
class MainActivity : ComponentActivity() {
    override fun onCreate(savedInstanceState: Bundle?) {
        super.onCreate(savedInstanceState)
        setContent {
            CleanAppTheme {
                Surface(color = MaterialTheme.
                    colors.background) {
                    val navController =
                        rememberNavController()
                    App(navController = navController)
                }
            }
        }
```

```
        }
      }
    }

@Composable
fun App(navController: NavHostController) {
    NavHost(navController, startDestination =
        "/posts") {
        composable(route = "/posts") {
            PostListScreen(hiltViewModel())
        }
    }
}
```

If we run the application, we should see the following screen:

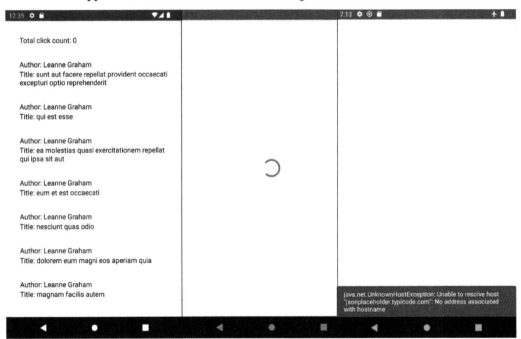

Figure 8.4 – Output of Exercise 08.01

We can see the list of post titles and the author's name for each post. The total click count is, for now, 0 because we haven't connected any logic and are yet to modify that value. We will add that logic in the exercises that follow. If an error occurs while loading this list, then we will see a snackbar with the description of the `Exception` object, and while the data is loaded, an indeterminate progress bar will be displayed.

In this section, we have implemented the presentation layer of an Android application using the MVVM architecture pattern and connected the layer to the domain layer of the application to display data to the user. In the section that follows, we will expand this layer across multiple modules and see how we can navigate between screens in different modules.

Presenting data in multiple modules

In this section, we will look at how we can separate the presentation layer into multiple modules, how we can handle the interaction between these modules, and how they can share the same data.

When developing Android applications, we can group screens into different modules. For example, we can group a login or registration flow inside a library module called *authentication*, or if we have a settings section, we can group those screens inside a separate module. Sometimes these screens will have commonalities with the rest of the application, such as using the same loading progress bar or the same error mechanism. Other times, these screens must navigate to screens from other modules. The question we now need to ask is how this can happen without creating a dependency between the two modules or other modules that are on the same level. Having a direct dependency on these modules will risk creating a cyclic dependency as shown here:

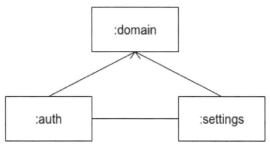

Figure 8.5 – Module cyclic dependency

In *Figure 8.5*, we show what might happen if we want to navigate from the :auth module to the :settings module and vice versa. This currently is impossible because of the cyclic dependency between the two modules. To solve this issue, we will need to create a new module. This module will hold the common logic shared between the two modules and common data. This will look like the following figure:

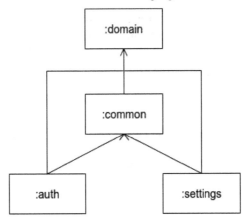

Figure 8.6 – Common presentation module

In *Figure 8.6*, we added the :common module, which will hold reusable views or @ Composable functions and the navigation data from the app. Over time, this module will grow, so it can be split into different modules holding different common features of the app (navigation, UI, common logic, and so on).

If we are using Jetpack Compose for our application, then we can rely on the work done in *Exercise 03.02 – Navigating using Jetpack Compose* of *Chapter 3, Understanding Data Presentation on Android*, where we defined the following structure for the app navigation:

```
private const val ROUTE_USERS = "users"
private const val ROUTE_USER = "users/%s"
private const val ARG_USER_NAME = "name"

sealed class AppNavigation(val route: String, val
    argumentName: String = "") {

    object Users : AppNavigation(ROUTE_USERS)

    object User : AppNavigation(String.format(ROUTE_USER,
        "{$ARG_USER_NAME}"), ARG_USER_NAME) {
```

```
        fun routeForName(name: String) =
            String.format(ROUTE_USER, name)
    }
}
```

The `routeForName` method was called from the `Users` screen when a user in the list was clicked, and then the `NavHost` method would use that route to open the `User` screen. When dealing with multiple modules, the routes that will be shared by the modules can be stored in the `:common` module so that each module will have access to the route. The `:app` module, which will have `NavHost`, will then be able to navigate between each screen.

When it comes to handling common logic between different modules, such as displaying the same error or loading views, we can declare the composable functions inside the `:common` module:

```
@Composable
fun Error(errorMessage: String) {
    ...
}

@Composable
fun Loading() {
    ...
}
```

If the same state is shared between the different screens in the different modules, we can have something like the following:

```
@Composable
fun <T> CommonScreen(state: State<T>, onSuccess:
    @Composable (T) -> Unit) {
    when (result) {
        is State.Success -> {
            onSuccess(result.data)
        }
        is State.Error -> {
            Error(result.errorMessage)
        }
```

```
        is State.Loading -> {
            Loading()
        }
    }
}
```

Here, we will check the current state and display the common error and loading views, leaving the screens themselves to only concern themselves with the successful state.

In this section, we have looked at how we can split the presentation layer into multiple modules and how to handle the common elements between these modules. In the following section, we will look at an exercise on how to achieve this. Splitting the presentation layer into multiple modules will decrease application build times because Gradle caching will only rebuild modules that contain changes. Another benefit comes in the form of drawing boundaries around the application's scope, which will be beneficial when it comes to exporting only certain features of an application.

Exercise 08.02 – Multi-module data presentation

Modify *Exercise 08.01 – Implementing MVVM* so that two new modules are created: `presentation-post` and `presentation-common`.

The `presentation-common` module will have the following:

- The `UiState` class, which will be moved from the `presentation-post` module.

- `CommonResultConverter`, which will be an abstract class with two methods: `convert`, which is a concrete method that will convert the `Result` object into a `UiState` object, and `convertSuccess`, which is an abstract method used to convert the data from `Result.Success`.

- `CommonScreen`, which will have the `@Composable` method for displaying the different types of `UiState` and two additional methods for displaying the error snackbar and the progress bar. The two methods will be moved from `PostListScreen`.

- `AppNavigation`, which will hold the routes to navigate to the list of posts, a single post, and a single user.

- The presentation-post module will have an additional package to display the information of a single post in the following format: Title: x and Body: y, where x is the title of a post and y is the body of the post. To display this information, a new ViewModel and Converter class will need to be created, which will convert the data from GetPostUseCase. When the author text is clicked, the app will navigate to the user screen, and when the Post list item is clicked, the app will navigate to the post screen. When either of these is clicked, UpdateInteractionUseCase is invoked to increase the number of clicks, which will then be reflected in the list header.

- presentation-user will display the information about a single user in the following format: Name: x, Username: y, and Email: z, where x, y, and z are represented by the information inside the User entity. The user data will be loaded from GetUserUseCase.

- The app module will be updated to handle the navigation between all these screens.

To complete this exercise, you will need to do the following:

1. Create the presentation-common module.

2. Move the UiState class and the Error and Loading @Composable functions and create a new @Composable function, which will handle each type of UiState object inside the CommonScreen file.

3. Create the CommonResultConverter class.

4. Create the AppNavigation class.

5. Modify the classes in presentation-post to reuse the preceding classes and methods.

6. Create the PostScreen, PostViewModel, PostConverter, and PostModel classes responsible for displaying the information about a single post.

7. Create the presentation-user module.

8. Create the UserScreen, UserViewModel, UserConverter, and UserModel classes responsible for displaying the information about a single post.

9. Implement the navigation between the screens.

10. Add the logic to update the number of clicks inside PostListViewModel.

Follow these steps to complete the exercise:

1. Create the presentation-common and presentation-user Android library modules.

2. Apply steps 3–5 from *Exercise 08.01 – Implementing MVVM* for each of these new modules.

3. In the `build.gradle` file of the `presentation-post` and `presentation-user` modules, make sure that the dependency to `presentation-common` is added:

```
dependencies {

    ...

    implementation(project(path: ":presentation-common"))

    ...

}
```

4. In the `presentation-common` module, create a new package called `state`.

5. Move the `UiState` class into the preceding package.

6. In the same package, create the `CommonResultConverter` class:

```
abstract class CommonResultConverter<T : Any, R : Any> {

    fun convert(result: Result<T>): UiState<R> {
        return when (result) {
            is Result.Error -> {
                UiState.Error(result.exception.
                    localizedMessage.orEmpty())
            }
            is Result.Success -> {
                UiState.Success(convertSuccess
                    (result.data))
            }
        }
    }

    abstract fun convertSuccess(data: T): R
}
```

Here, we return `UiState.Error` for any `Result.Error` object with the exception message, and for `Result.Success`, we return `UiState.Success` and use an abstraction for the data inside the `Result.Success` object. This represents a solution for how we can extract the common logic for displaying the error.

7. Modify the `PostListConverter` class from the `presentation-post` module
    so that it will extend `CommonResultConverter` and provide an implementation
    for the `convertSuccess` method:

```
class PostListConverter @Inject constructor(@
ApplicationContext private val context: Context) :
    CommonResultConverter<GetPostsWithUsersWithInteraction
    UseCase.Response, PostListModel>() {

    override fun convertSuccess(data:
        GetPostsWithUsersWithInteractionUseCase.
            Response): PostListModel {
        return PostListModel(
            headerText = context.getString(
                R.string.total_click_count,
                data.interaction.totalClicks
            ),
            items = data.posts.map {
                PostListItemModel(
                    it.post.id,
                    it.user.id,
                    context.getString(R.string.author,
                        it.user.name),
                    context.getString(R.string.title,
                        it.post.title)
                )
            }
        )
    }
}
```

Here, we only deal with converting
`GetPostsWithUsersWithInteractionUseCase.Response` into
`PostListModel`, allowing the parent class to handle the error only.

8. In the `state` package from the `presentation-common` module, create a new
    file called `CommonScreen`.

9. In the `CommonScreen` file, add a `CommonScreen` `@Composable` method, which
    will check `UiState` and invoke `Error` for `UiState.Error` and `Loading` for
    `UiState.Loading`:

```
@Composable
fun <T : Any> CommonScreen(state: UiState<T>, onSuccess:
@Composable (T) -> Unit) {
    when (state) {
        is UiState.Loading -> {
            Loading()
        }
        is UiState.Error -> {
            Error(errorMessage = state.errorMessage)
        }
        is UiState.Success -> {
            onSuccess(state.data)
        }
    }
}
```

10. Move the `Error` and `Loading` `@Composable` functions from
    `PostListScreen` into the `CommonScreen` file.

11. Modify the `PostListScreen` `@Composable` method from the
    `presentation-post` module so that it will use the `CommonScreen` method:

```
@Composable
fun PostListScreen(
    viewModel: PostListViewModel
) {
    viewModel.loadPosts()
    viewModel.postListFlow.collectAsState().value.let
        { state ->
        CommonScreen(state = state) {
            PostList(postListModel = it)
        }
    }
}
```

Now the entire logic for converting and showing the list of posts will only deal with the associated objects, leaving the error and loading scenarios in the `presentation-common` module.

12. In `presentation-common`, create a new package called `navigation`.

13. In the `navigation` package, create a class called `PostInput`:

```
data class PostInput(val postId: Long)
```

This class is meant to represent the input that the post screen will require to load its data.

14. In the same package, create a class called `UserInput`:

```
data class UserInput(val userId: Long)
```

This class is meant to represent the input that the user screen will require to load its data.

15. In the same package, create a new class called `NavRoutes`:

```
private const val ROUTE_POSTS = "posts"
private const val ROUTE_POST = "posts/%s"
private const val ROUTE_USER = "users/%s"
private const val ARG_POST_ID = "postId"
private const val ARG_USER_ID = "userId"


sealed class NavRoutes(
    val route: String,
    val arguments: List<NamedNavArgument> =
        emptyList()
) {

    ...

}
```

Here, we define the paths for each screen. The posts screen will have no arguments, but the user and post screens will require the `postId` and `userId` values.

16. Create the `Posts` class in the `NavRoutes` class:

```
sealed class NavRoutes(
    val route: String,
    val arguments: List<NamedNavArgument> =
        emptyList()
```

```
) {
    object Posts : NavRoutes(ROUTE_POSTS)
}
```

17. Create the `Post` class in the `NavRoutes` class:

```
sealed class NavRoutes(
    val route: String,
    val arguments: List<NamedNavArgument> =
        emptyList()
) {
    object Post : NavRoutes(
        route = String.format(ROUTE_POST,
            "{$ARG_POST_ID}"),
        arguments = listOf(navArgument(ARG_POST_ID) {
            type = NavType.LongType
        })
    ) {

        fun routeForPost(postInput: PostInput) =
            String.format(ROUTE_POST, postInput.postId)

        fun fromEntry(entry: NavBackStackEntry):
            PostInput {
            return PostInput(entry.arguments?.
                getLong(ARG_POST_ID) ?: 0L)
        }
    }
}
```

Here, we will need to break down the `Post` input into the arguments for the URL. The `routeForPost` method will create a `/posts/1` URL for a `Post` object that has the ID `1`. The `fromEntry` method will re-assemble the `PostInput` object from the navigation entry object. The reason we are taking this approach is that the navigation library discourages the use of `Parcelable`, which means that passing data between different screens will have to be done through the URL. To avoid any issues with keeping track of the arguments across multiple modules, we can instead use objects and keep the logic to read from arguments and construct the arguments isolated to this class.

18. Create the `User` class inside the `NavRoutes` class:

```
sealed class NavRoutes(
    val route: String,
    val arguments: List<NamedNavArgument> = emptyList()
) {
    object User : NavRoutes(
        route = String.format(ROUTE_USER,
            "{$ARG_USER_ID}"),
        arguments = listOf(navArgument(ARG_USER_ID) {
            type = NavType.LongType
        })
    ) {

        fun routeForUser(userInput: UserInput) =
            String.format(ROUTE_USER, userInput.userId)

        fun fromEntry(entry: NavBackStackEntry):
            UserInput {
            return UserInput(entry.arguments?.getLong
                (ARG_USER_ID) ?: 0L)
        }
    }
}
```

Here, we apply the same principle as we did for the `Post` class.

19. Create a new package called `single` in the `presentation-post` module.

20. In the `single` package, create the `PostModel` class:

```
data class PostModel(
    val title: String,
    val body: String
)
```

21. In the `single` package, create the `PostConverter` class:

```
class PostConverter @Inject constructor(@
ApplicationContext private val context: Context) :
    CommonResultConverter<GetPostUseCase.Response,
```

```
        PostModel>() {

    override fun convertSuccess(data:
        GetPostUseCase.Response): PostModel {
        return PostModel(
            context.getString(R.string.title,
                data.post.title),
            context.getString(R.string.body,
                data.post.body)
        )
    }
}
```

22. Add the body string to `strings.xml` of the `presentation-post` module:

```
<resources>
    …
    <string name="body">Body: %s</string>
</resources>
```

23. In the `single` package, create the `PostViewModel` class:

```
@HiltViewModel
class PostViewModel @Inject constructor(
    private val postUseCase: GetPostUseCase,
    private val postConverter: PostConverter
) : ViewModel() {

    private val _postFlow =
        MutableStateFlow<UiState<PostModel>>(UiState.
Loading)
    val postFlow: StateFlow<UiState<PostModel>> =
        _postFlow

    fun loadPost(postId: Long) {
        viewModelScope.launch {
            postUseCase.execute(GetPostUseCase.
                Request(postId))
```

```
            .map {
                postConverter.convert(it)
            }
            .collect {
                _postFlow.value = it
            }
        }
    }
}
```

Here, we are using GetPostUseCase to load the information about a particular post and are using the converter defined earlier to convert the data into PostModel, which will be set in the Flow object.

24. In the single package, create the PostScreen file, which will display the post information:

```
@Composable
fun PostScreen(
    viewModel: PostViewModel,
    postInput: PostInput
) {
    viewModel.loadPost(postInput.postId)
    viewModel.postFlow.collectAsState().value.let {
        result ->
        CommonScreen(result) { postModel ->
            Post(postModel)
        }
    }
}
```

```
@Composable
fun Post(postModel: PostModel) {
    Column(modifier = Modifier.padding(16.dp)) {
        Text(text = postModel.title)
        Text(text = postModel.body)
    }
}
```

Here, we follow the same principle as for the `PostListScreen` file, where we split into two methods, `PostScreen` for observing the `UiState` object and `PostListScreen` to deal with drawing the user interface.

25. In the `presentation-user` module, create a new package called `single`.

26. In the `single` package, create a new class called `UserModel`:

```
data class UserModel(
    val name: String,
    val username: String,
    val email: String
)
```

27. In the `single` package, create a new class called `UserConverter`:

```
class UserConverter @Inject constructor(@
ApplicationContext private val context: Context) :
    CommonResultConverter<GetUserUseCase.Response,
        UserModel>() {

    override fun convertSuccess(data: GetUserUseCase.
        Response): UserModel {
        return UserModel(
            context.getString(R.string.name,
                data.user.name),
            context.getString(R.string.username,
                data.user.username),
            context.getString(R.string.email,
                data.user.email)
        )
    }
}
```

28. Create the `res/values/strings.xml` file inside the `main` folder in the `presentation-user` module:

```
<?xml version="1.0" encoding="utf-8"?>
<resources>
    <string name="name">Name: %s</string>
    <string name="username">Username: %s</string>
```

```
        <string name="email">Email: %s</string>
    </resources>
```

29. Inside the `single` package, create `UserViewModel`:

```
@HiltViewModel
class UserViewModel @Inject constructor(
    private val userUseCase: GetUserUseCase,
    private val converter: UserConverter
) : ViewModel() {

    private val _userFlow =
        MutableStateFlow<UiState<UserModel>>
            (UiState.Loading)
    val userFlow: StateFlow<UiState<UserModel>> =
        _userFlow

    fun loadUser(userId: Long) {
        viewModelScope.launch {
            userUseCase.execute
                (GetUserUseCase.Request(userId))
                .map {
                    converter.convert(it)
                }
                .collect {
                    _userFlow.value = it
                }
        }
    }
}
```

Here, we take the user data from `GetUserUseCase`, convert it using `UserConverter`, and post the result in the `Flow` object.

30. In the `single` package, create the `UserScreen` file:

```
@Composable
fun UserScreen(
    viewModel: UserViewModel,
```

```
        userInput: UserInput
) {
    viewModel.loadUser(userInput.userId)
    viewModel.userFlow.collectAsState().value.let {
        result ->
        CommonScreen(result) { userModel ->
            User(userModel)
        }
    }
}


@Composable
fun User(userModel: UserModel) {
    Column(modifier = Modifier.padding(16.dp)) {
        Text(text = userModel.name)
        Text(text = userModel.username)
        Text(text = userModel.email)
    }
}
```

Here, we take the same approach as the other screens, where in one method, we subscribe to changes in `UiState`, and in the other, we display the user information.

31. Add the click listeners in `PostListScreen`:

```
@Composable
fun PostList(
    postListModel: PostListModel,
    onRowClick: (PostListItemModel) -> Unit,
    onAuthorClick: (PostListItemModel) -> Unit
) {
    LazyColumn(modifier = Modifier.padding(16.dp)) {
        ...
        items(postListModel.items) { item ->
            Column(modifier = Modifier
                .padding(16.dp)
                .clickable {
                    onRowClick(item)
```

```
        }) {
        ClickableText(text = AnnotatedString(
            text = item.authorName), onClick =
        {
            onAuthorClick(item)
        })
        Text(text = item.title)
    }
        }
    }
}
```

In the preceding snippet, we specify click listeners for when the row is clicked and for when the author is clicked. Because we are applying state hoisting, we want to propagate the click listeners to the caller of the PostList method. To achieve this, we define a parameter for each click listener as a lambda function that has as input the row data and requires no result. More information about lambdas can be found here: https://kotlinlang.org/docs/lambdas.html#function-types.

32. Modify the PostListScreen @Composable method so that when the user is clicked, we navigate to the user screen, and when the row is clicked, we navigate to the post:

```
@Composable
fun PostListScreen(
    viewModel: PostListViewModel,
    navController: NavController
) {
    viewModel.loadPosts()
    viewModel.postListFlow.collectAsState().value.let
        { state ->
        CommonScreen(state = state) {
            PostList(it, { postListItem ->
                navController.navigate(NavRoutes.Post.
                    routeForPost(PostInput
                        (postListItem.id)))
            }) { postListItem ->
                navController.navigate(NavRoutes.User.
```

```
                        routeForUser(UserInput
                            (postListItem.userId)))
                }
            }
        }
    }
```

33. In `build.gradle` of the app module, make sure that the dependencies to `presentation-common` and `presentation-user` are added:

```
dependencies {
    ...
    implementation(project(path: ":presentation-
        user"))
    implementation(project(path: ":presentation-
        common"))
    ...
}
```

34. In the `MainActivity` file, modify the `App` method so that the navigation between the different screens is implemented:

```
@Composable
fun App(navController: NavHostController) {
    NavHost(navController, startDestination =
        NavRoutes.Posts.route) {
        composable(route = NavRoutes.Posts.route) {
            PostListScreen(hiltViewModel(),
                navController)
        }
        composable(
            route = NavRoutes.Post.route,
            arguments = NavRoutes.Post.arguments
        ) {
            PostScreen(
                hiltViewModel(),
                NavRoutes.Post.fromEntry(it)
            )
```

```
            }
        composable(
            route = NavRoutes.User.route,
            arguments = NavRoutes.User.arguments
        ) {
            UserScreen(
                hiltViewModel(),
                NavRoutes.User.fromEntry(it)
            )
        }
    }
}
```

Here, we add all the screens in the application to the navigation graph, and in the case of UserScreen and PostScreen, we extract the UserInput and PostInput objects from the navigation graph entries. We will now need to add the interaction.

35. Add an Interaction field inside PostListModel:

```
data class PostListModel(
    …
    val interaction: Interaction
)
```

36. Modify PostListConverter to include the interaction field:

```
class PostListConverter @Inject constructor(@
ApplicationContext private val context: Context) :
    CommonResultConverter<GetPostsWithUsersWithInteraction
    UseCase.Response, PostListModel>() {

    override fun convertSuccess(data:
        GetPostsWithUsersWithInteractionUseCase.
            Response): PostListModel {
        return PostListModel(
            …
            interaction = data.interaction
        )
```

```
        }
    }
```

37. Add a reference to `UpdateInteractionUseCase` in `PostListViewModel` and a method to update the interaction:

```
@HiltViewModel
class PostListViewModel @Inject constructor(
    ...
    private val updateInteractionUseCase:
        UpdateInteractionUseCase
) : ViewModel() {
    ...
    fun updateInteraction(interaction: Interaction) {
        viewModelScope.launch {
            updateInteractionUseCase.execute(
                UpdateInteractionUseCase.Request(
                    interaction.copy(
                        totalClicks = interaction.
                            totalClicks + 1
                    )
                )
            ).collect()
        }
    }
}
```

38. Modify the `PostListScreen` @Composable method so that it will call to update the interaction for each click:

```
@Composable
fun PostListScreen(
    viewModel: PostListViewModel,
    navController: NavController
) {
    ...
    viewModel.postListFlow.collectAsState().value.let
        { state ->
```

```
CommonScreen(state = state) {
    PostList(it, { postListItem ->
        viewModel.updateInteraction(it.
interaction)

        ...

    }) { postListItem ->
        viewModel.updateInteraction(it.
interaction)

        ...

    }
        }
    }
}
```

If we run the application, we will see an output like the one in the following figure:

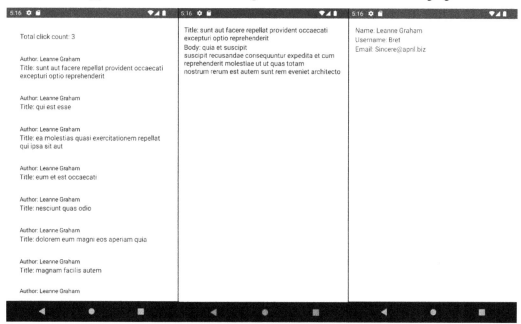

Figure 8.7 – Output of Exercise 08.02

We can see from *Figure 8.7* that when the row is clicked, we are taken to the screen displaying the post information, and when the author is clicked, we are taken to the user information. By placing the `NavRoutes` class in the `presentation-common` module, we can navigate from the post list on a screen located in the same module (post) and a screen located in a different module (user). The solution of creating additional modules is a good way to avoid cyclic dependencies not only for modules in the presentation layer but also for modules in the other layers as well.

In this exercise, we have learned how to split the presentation layer into separate modules and how we can use a common module to hold shared logic and data required by all the modules in the layer. This is a technique that can be used for other layers in the application if we want them split up as well.

Summary

In this chapter, we explored the presentation layer of an Android application and a few different approaches for implementing this layer, such as MVC, MVP, and MVVM. We decided to focus on the MVVM approach because of the many benefits involving the life cycle and the compatibility with Jetpack Compose. We then looked at what happens when we want to split the presentation layer across multiple modules and how we can solve the common logic between these modules. In the chapter that follows, we will further build upon the MVVM pattern and study the **Model-View-Intent** (**MVI**) pattern, which further takes advantage of the Observable pattern to incorporate the user actions into states that can be observed.

9

Implementing an MVI Architecture

In this chapter, we will introduce the concept of **Model-View-Intent** (**MVI**) and the benefits it provides for managing the state of an application. We will begin by analyzing what MVI is and then move on to implementing it using Kotlin flows. In this chapter's exercise, we will build upon the previous chapter's exercises, and we will re-implement them using the MVI pattern to highlight how this pattern can be integrated into the presentation layer of an application with multiple modules.

In this chapter, we will cover the following topics:

- Introducing MVI
- Implementing MVI with Kotlin flows

By the end of the chapter, you will be able to implement the MVI architecture pattern inside a multimodule Android application, using Kotlin flows.

Technical requirements

The hardware and software requirements are as follows:

- Android Studio Arctic Fox 2020.3.1 patch 3

The code files for this chapter can be found here: `https://github.com/PacktPublishing/Clean-Android-Architecture/tree/main/Chapter9`.

Check out the following video to see the Code in Action: `https://bit.ly/3FYZKLn`

Introducing MVI

In this section, we will look at what the MVI architecture pattern is, the problems it is trying to solve, and the solutions it presents for solving those problems.

Let's imagine you need to develop a configuration screen for an application. It will load the existing configuration and it will need to toggle various switches and prepopulate input fields with the existing data. After that data is loaded, then the user can modify each of those fields. To achieve this, you would probably need to keep mutable references for the data represented in those fields so that when the user changes a value, the reference changes.

This may pose a problem because of the mutability of those fields, especially when dealing with concurrent operations or their order. A solution to this problem is to make the data immutable and combine it into a state that the user interface can observe. Any changes the app or user will need to make on the user interfaces will be through a reactive data flow. The flow will then create a new state representing the change and update the user interface.

This is essentially how MVI operates. In MVI, the **View** plays the same role as in MVP or MVVM and the **Model** holds the state of the user interface, and it represents the single source of truth. The **Intent** is represented by any changes that should be made to the state, which will then be updated. In *Figure 9.1*, we can see how the **View** will send an **Intent** to the **Model**, which will then trigger a change in state, which will update the **View**:

> **Note**
> The term Intent in the context of MVI is different from the Android `Intent` class used to interact with different Android components.

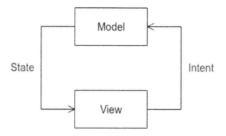

Figure 9.1 – MVI diagram

What is missing from this diagram is the equivalent of a **ViewModel** or a **Presenter**. This is because the MVI pattern isn't a replacement for those patterns but instead builds on top of them.

To visualize how this might look, let's look at an example of a `ViewModel`:

```
class MyViewModel @Inject constructor(
    private val getMyDataUseCase: GetMyDataUseCase
) : ViewModel() {
    private val _myDataFlow =
        MutableStateFlow<MyData>(MyData())
    val myDataFlow: StateFlow<MyData> = _myDataFlow


    var text: String = ""


    fun loadMyData() {
        viewModelScope.launch {
            getMyDataUseCase.execute
                (GetPostsWithUsersWithInteractionUseCase.
                    Request)
                .collect {
                    _myDataFlow.value = it
                }
        }
    }
}
```

In the preceding example, we define a class named `MyViewModel` in which we have a use case to load data and a `text` variable that will be changed by the View when the user changes it. We can see that the `text` variable is a mutable variable accessible from the **View**. We also have a `StateFlow` variable holding the data to be loaded and we have a method to load the data. To transition the preceding code to MVI, we will need to first define a state that will hold the data to be loaded and the text. This will represent our source of truth. For the preceding example, this state will look as in the following example:

```
data class MyState(
    val myData: MyData = MyData(),
    val text: String = ""
)
```

In the `MyState` class, we move the data to be loaded and the text to be changed. Now, we will need to identify the actions; in this case, we have two actions: loading the data and updating the value of the text with a new value introduced by the user:

```
sealed class MyAction {
    object LoadAction : MyAction()
    data class UpdateAction(val text: String) :
        MyAction()
}
```

In the preceding example, we have represented the action as a sealed class and defined two actions for loading and updating the text. Next, we will need to create the appropriate data flows for handling the actions and managing the state:

```
private val _myStateFlow = MutableStateFlow<MyState>
    (MyState())
val myStateFlow: StateFlow<MyState> = _myDataFlow
private val _actionFlow: MutableSharedFlow<MyAction> =
    MutableSharedFlow()
```

In the preceding example, we have changed the `StateFlow` variables to hold the state object defined previously and added a similar `SharedFlow` variable, which will be responsible for managing the actions inserted by the user. We will now need to subscribe and handle the actions:

```
class MyViewModel @Inject constructor(
    private val getMyDataUseCase: GetMyDataUseCase
) : ViewModel() {
```

```
...
    init {
        viewModelScope.launch {
            action.collect { action ->
                when (action) {
                    is MyViewModel.MyAction.LoadAction -> {
                        loadMyData()
                    }
                    is MyViewModel.MyAction.UpdateAction -> {
                        _myDataFlow.emit(_myDataFlow.value.
copy(text =
    action.text))
                    }
                }
            }

        }
    }
}


    fun submitAction(action: MyAction) {
        viewModelScope.launch {
            _action.emit(action)
        }
    }



    private fun loadMyData() {
        getMyDataUseCase.execute
            (GetPostsWithUsersWithInteractionUseCase.
                Request)
            .collect {
                _myDataFlow.value = it
            }
    }
    ...
}
```

In the `init` block, we are collecting the actions and then, for each action, we perform the required operation. The View will invoke the `submitAction` method and pass the action it wants the ViewModel to perform. For this example, `MyAction` plays the role of the Intent within the MVI context and the ViewModel will sit between the View and Model and will be responsible for managing the flow of data between the Model and the View, as well as managing the state.

When it comes to the implementation of the MVI pattern, there are many different variations for different technologies and different architecture patterns. From RxJava to `LiveData`, to flows and coroutines, to MVVM and MVP, there are different approaches to the pattern with different variations.

Some are built using concepts such as state machines, others use basic streams, and others use third-party open source libraries. From the preceding example, we can see that the pattern introduces a little bit of boilerplate code, so it is important to perform research and monitor the initial introduction of the pattern into any application. In the section that follows, we will look at how we can implement MVI using Kotlin flows.

Implementing MVI with Kotlin flows

In this section, we will look at how we can implement the MVI architecture pattern using Kotlin flows and the benefits and pitfalls of this approach.

In the previous section, we defined an MVI approach using `StateFlow` and `SharedFlow`, as in the following example:

```
    private val _myStateFlow =
MutableStateFlow<MyState>(MyState())
    val myStateFlow: StateFlow<MyState> = _myDataFlow
    private val actionFlow: MutableSharedFlow<MyAction> =
MutableSharedFlow()
```

The different types of flows used here serve different purposes. `MutableStateFlow` will emit the last value held, which is good for the user interface because we want it to display the last data loaded, like how `LiveData` works. `SharedFlow` doesn't have this feature, which is useful for actions because we do not want the last action to be emitted twice. Another aspect we will need to consider is one-shot events, which should be emitted using a channel flow. This will be useful when the View will need to respond to events in a channel to display a toast alert or handle navigation to a new screen. We can apply this using the following:

```
class MyViewModel @Inject constructor(
```

```kotlin
    private val getMyDataUseCase: GetMyDataUseCase
) : ViewModel() {

    ...

    private val _myStateFlow = MutableStateFlow<MyState>
        (MyState())
    val myStateFlow: StateFlow<MyState> = _myDataFlow
    private val actionFlow: MutableSharedFlow<MyAction> =
        MutableSharedFlow()
    private val _myOneOffFlow = Channel<MyOneOffEvent>()
    val myOneOffFlow = _myOneOffFlow.receiveAsFlow()

    ...

}
```

In the preceding example, we have integrated the `Channel` information with the rest of the `ViewModel`. Because an application will end up having multiple ViewModels, we can create a template that will be used across the application. We can start by defining abstractions for each of the state, action, and one-off events:

```kotlin
interface UiState
```

```kotlin
interface UiAction
```

```kotlin
interface UiSingleEvent
```

Here, we have opted for a simple interface to represent each of the flows of data the `ViewModel` will use. We can next define a template for the `ViewModel`, which can be inherited by the ViewModels used in the application:

```kotlin
abstract class MviViewModel<S : UiState, A : UiAction, E :
UiSingleEvent> : ViewModel() {

    private val _uiStateFlow: MutableStateFlow<S> by lazy {
        MutableStateFlow(initState())
    }
    val uiStateFlow: StateFlow<S> = _uiStateFlow
    private val actionFlow: MutableSharedFlow<A> =
        MutableSharedFlow()
    private val _singleEventFlow = Channel<E>()
    val singleEventFlow = _singleEventFlow.receiveAsFlow()
```

```
    ...
}
```

In the preceding example, we have used generics for each of the flows that the
`ViewModel` will use. This creates a problem for `MutableStateFlow`, which requires
an initial value. Because we don't have any concrete value to initialize, we will need to
create an abstract method that will provide the initial value:

```
abstract class MviViewModel<S : UiState, A : UiAction, E :
UiSingleEvent> : ViewModel() {

    ...

    init {
        viewModelScope.launch {
            actionFlow.collect {
                handleAction(it)
            }
        }
    }

    abstract fun initState(): S

    abstract fun handleAction(action: A)

}
```

In addition to the `initState` abstraction, we have also added the `handleAction`
abstraction. This will be called when new actions are submitted because of user actions
or a screen load. Because the mutable variables are set to private, we will need to expose
methods that emit events into these flows:

```
abstract class MviViewModel<S : UiState, A : UiAction, E :
    UiSingleEvent> : ViewModel() {

    ...

    fun submitAction(action: A) {
        viewModelScope.launch {
            actionFlow.emit(action)
        }
    }
```

```kotlin
fun submitState(state: S) {
    viewModelScope.launch {
        _uiStateFlow.value = state
    }
}

fun submitSingleEvent(event: E) {
    viewModelScope.launch {
        _singleEventFlow.send(event)
    }
}
}
```

In the preceding example, we have added the methods that emit, send, or change the value on each of the specific data flows. To implement the template for a specific scenario, we will need to create concretions for UiState:

```kotlin
sealed class MyUiState : UiState {
    data class Success(val myData: MyData) : MyUiState()
    object Error : MyUiState()
    object Loading : MyUiState()
}
```

In the preceding example, we have defined different states that the screen might have. We can now create a concretion for UiAction:

```kotlin
sealed class MyUiAction : UiAction {
    object Load : MyUiAction()
    object Click : MyUiAction()
}
```

Here, we defined an action for when the data will need to be loaded and another for when something is clicked on the user interface:

```kotlin
sealed class MyUiSingleEvent : UiSingleEvent {
    data class ShowToast(val text: String) :
        MyUiSingleEvent()
}
```

For the single event fired, we have defined a show toast alert event. Finally, we can implement the concretion for the `ViewModel`:

```
class MyViewModel : MviViewModel<MyUiState, MyUiAction,
    MyUiSingleEvent>() {


    override fun initState(): MyUiState = MyUiState.Loading


    override fun handleAction(action: MyUiAction) {
        when (action) {
            is MyUiAction.Load -> {
                viewModelScope.launch {
                    val state: UiState = // Fetch UI state
                    submitState(state)
                }
            }
            is MyUiAction.Click -> {
                // Handle logic for clicks
                submitSingleEvent(MyUiSingleEvent.
                    ShowToast("Toast"))
            }
        }
    }
}
```

In the preceding example, we have extended the `MviViewModel` class and passed `MyUiState`, `MyUiAction`, and `MyUiSingleEvent` for the generics. In the `initState` method, we return the `Loading` state, and in the `handleAction` method, we check the actions and then load the data or handle the click event, which will then submit the event to show a toast alert.

If we want to integrate the `ViewModel` with Jetpack Compose, we will have to use something like the following example:

```
@Composable
fun MyScreen(
    viewModel: MyViewModel
) {
    viewModel.submitAction(MyUiAction.Load)
```

```
viewModel.uiStateFlow.collectAsState().value.let {
    state ->
    when (state) {
        is MyUiState.Loading -> {

        }
        is MyUiState.Success -> {
            MySuccessScreen(state.myData) {
                viewModel.submitAction(MyUiAction.
                    Click)
            }
        }
        is MyUiState.Error -> {

        }
    }
}
```

We can see that observing UiState will remain the same as for MVVM; however, if we wish to notify the ViewModel of any changes, we will need to use the submitAction method. For the UiSingleEvents object, we will need to use the LaunchedEffect function because we don't want Jetpack Compose to keep recomposing and re-executing the same block; we only want it to be executed once, so we will need to use something such as the following:

```
@Composable
fun MyScreen(
    viewModel: MyViewModel
) {

    ...

    LaunchedEffect(Unit, {
        viewModel.singleEventFlow.collectLatest {
            when (it) {
                is MyUiSingleEvent.ShowToast -> {
                    // Show Toast
                }
```

```
            }
        }
    })
}
```

In this example, we collect the data from `Channel` inside the `LaunchedEffect` method and then show a toast alert when the `ShowToast` event is received. `LaunchedEffect` can also be used to ensure that we do not trigger multiple data loads because of the Jetpack Compose recomposition mechanism:

```
@Composable
fun MyScreen(
    viewModel: MyViewModel
) {
    LaunchedEffect(Unit, {
        viewModel.submitAction(MyUiAction.Load)
    }
}
```

In the preceding snippet, we have moved the call to `submitAction` inside `LaunchedEffect`, to avoid triggering the loading multiple times. More information about Jetpack Compose side effects can be found here: `https://developer.android.com/jetpack/compose/side-effects`.

In this section, we have shown how we can integrate the MVI architecture pattern with flows and Jetpack Compose. We have seen how we have translated the interactions between the View and the `ViewModel` into intents using the `UiAction` interface and the implementations of this interface. We have also seen some of the downsides of the pattern because of the addition of boilerplate code and, in the case of Jetpack Compose, having to use methods such as `LaunchedEffect` and `Channel` for emitting one-off events. In the following section, we will create an application in which we will migrate a previous exercise to use MVI.

Exercise 09.01 – Transitioning to MVI

Modify *Exercise 08.02 – Multi-module data presentation* from *Chapter 8, Implementing an MVVM Architecture*, so that the presentation layer uses the MVI architecture pattern. The UiState class will remain and represent the state of each screen. New interfaces will be added in the presentation-common module representing actions and one-off events. In the same module, an MviViewModel abstract class will be implemented, which will be the template for the other ViewModels used in the application. For PostListViewModel, we will create new user actions for loading the data, clicking on the post, and clicking on the author, and two new one-off events will be needed for opening each of those screens. For PostViewModel and UserViewModel, we will create only a single user action, which will be responsible for loading the data on each screen.

To complete this exercise, you will need to do the following:

1. In presentation-common, create an interface called UiAction and an interface called UiSingleEvent, and then create the MviViewModel template.

2. In the list package of the presentation-post module, create a sealed class called PostListUiAction, which will contain three subclasses called Load, UserClick, and PostClick. Then, create a sealed class called PostListUiSingleEvent, which will have two subclasses named OpenUserScreen and OpenPostScreen. Then, modify PostListViewModel and PostListScreen to use the specified actions and events.

3. In the single package of the presentation-post module, create a sealed class called PostUiAction, which will have one subclass named Load, which will contain the ID of the post. Then, modify PostViewModel and PostScreen to instead use the specified action.

4. In the single package of the presentation-user module, create a sealed class called UserUiAction, which will have one subclass named Load, which will contain the ID of the user. Then, modify UserViewModel and UserScreen to instead use the specified action.

Follow these steps to complete the exercise:

1. In the state package of the `presentation-common` module, create an interface called `UiAction`:

```
interface UiAction
```

2. In the same package, create an interface called `UiSingleEvent`:

```
interface UiSingleEvent
```

3. In the same package, create an abstract class called `MviViewModel`:

```
abstract class MviViewModel<T : Any, S : UiState<T>, A :
UiAction, E : UiSingleEvent> : ViewModel() {

}
```

Because we are using the `UiState` class with generics, we will need to also supply that generic field in the generic specification of `MviViewModel`.

4. In the `MviViewModel` class, add the necessary flows and channels that will hold the states, actions, and events:

```
abstract class MviViewModel<T : Any, S : UiState<T>, A :
UiAction, E : UiSingleEvent> : ViewModel() {


    private val _uiStateFlow: MutableStateFlow<S> by
        lazy {
        MutableStateFlow(initState())
    }
    val uiStateFlow: StateFlow<S> = _uiStateFlow
    private val actionFlow: MutableSharedFlow<A> =
        MutableSharedFlow()
    private val _singleEventFlow = Channel<E>()
    val singleEventFlow = _singleEventFlow.
        receiveAsFlow()
}
```

In this snippet, we have defined `StateFlow` variables to hold the last value that was emitted, which will be used to manage the state of the user interface, `SharedFlow`, which is used for handling user actions, and `Channel` for handling emitting one-off events. In the `MviViewModel` class, we are also defining generics so that we bind states, actions, and one-off events to their respective types.

5. In `MviViewModel`, add the abstract methods for initializing the state and handling the actions:

```
abstract class MviViewModel<T : Any, S : UiState<T>, A :
UiAction, E : UiSingleEvent> : ViewModel() {

    ...

    init {
        viewModelScope.launch {
            actionFlow.collect {
                handleAction(it)
            }
        }
    }

    abstract fun initState(): S

    abstract fun handleAction(action: A)
}
```

In this snippet, we are adding the abstraction required to provide an initial value for `StateFlow`, and then we handle the collection of the user actions, which will be handled in the `handleAction` method.

6. In `MviViewModel`, add the required methods to submit the state, events, and actions:

```
abstract class MviViewModel<T : Any, S : UiState<T>, A :
UiAction, E : UiSingleEvent> : ViewModel() {

    ...

    fun submitAction(action: A) {
        viewModelScope.launch {
            actionFlow.emit(action)
        }
    }

    fun submitState(state: S) {
        viewModelScope.launch {
            _uiStateFlow.value = state
        }
```

```
        }

        fun submitSingleEvent(event: E) {
            viewModelScope.launch {
                _singleEventFlow.send(event)
            }
        }
    }
```

In this snippet, we are defining a set of methods to emit data into the two `Flow` objects and the `Channel` object.

7. In the `list` package of the `presentation-post` module, create the `PostListUiAction` class and its subclasses:

```
sealed class PostListUiAction : UiAction {
    object Load : PostListUiAction()
    data class UserClick(val userId: Long, val
        interaction: Interaction) : PostListUiAction()
    data class PostClick(val postId: Long, val
        interaction: Interaction) : PostListUiAction()
}
```

Here, we define a sealed class for loading the data and clicking on the user and the post. Each of them will implement the `UiAction` interface.

8. In the same package, create the `PostListUiAction` class and its subclasses:

```
sealed class PostListUiSingleEvent : UiSingleEvent {
    data class OpenUserScreen(val navRoute: String) :
        PostListUiSingleEvent()
    data class OpenPostScreen(val navRoute: String) :
        PostListUiSingleEvent()
}
```

Here, we define a sealed class for the one-off events that will be emitted when we want the user and post screens to be opened, which is why we are implementing `UiSingleEvent`.

9. In the same package, modify `PostListViewModel` to extend `MviViewModel`:

```
@HiltViewModel
```

```
class PostListViewModel @Inject constructor(
    private val useCase:
        GetPostsWithUsersWithInteractionUseCase,
    private val converter: PostListConverter,
    private val updateInteractionUseCase:
        UpdateInteractionUseCase
) : MviViewModel<PostListModel, UiState<PostListModel>
    , PostListUiAction, PostListUiSingleEvent>() {

    ...

}
```

In this snippet, we are extending MviViewModel and providing the types we
defined previously as well as the existing PostListModel type to the generic
fields. This is because we want this ViewModel to be bound to the data, actions,
and one-off events that occur in PostListScreen.

10. Implement the initState method in the PostListViewModel class:

```
@HiltViewModel
class PostListViewModel @Inject constructor(
    ...
) : MviViewModel<PostListModel, UiState<PostListModel>
    , PostListUiAction, PostListUiSingleEvent>() {

    override fun initState(): UiState<PostListModel> =
        UiState.Loading
}
```

In this snippet, we are implementing the initState method and providing the
UiState.Loading value, which will in turn make the uiStateFlow field from
the parent class be initialized with the Loading value.

11. Implement the handleAction method in the PostListViewModel class:

```
@HiltViewModel
class PostListViewModel @Inject constructor(
    ...
) : MviViewModel<PostListModel, UiState<PostListModel>
    , PostListUiAction, PostListUiSingleEvent>() {

    ...
```

```
override fun handleAction(action:
    PostListUiAction) {
    when (action) {
        is PostListUiAction.Load -> {
            loadPosts()
        }
        is PostListUiAction.PostClick -> {
            updateInteraction(action.interaction)
            submitSingleEvent(
                PostListUiSingleEvent.
                    OpenPostScreen(
                    NavRoutes.Post.routeForPost(
                        PostInput(action.postId)
                    )
                )
            )
        }
        is PostListUiAction.UserClick -> {
            updateInteraction(action.interaction)
            submitSingleEvent(
                PostListUiSingleEvent.
                    OpenUserScreen(
                    NavRoutes.User.routeForUser(
                        UserInput(action.userId)
                    )
                )
            )
        }
    }
}
```

In this snippet, we are implementing the handleAction method, which will check what action we will need to handle and perform the necessary operation for each. For loading, we will invoke the loadPosts method, and for clicking on a user and a post, we will invoke the updateInteraction method and then submit a one-off event to open the user and post screens.

12. Implement the `loadPosts` method in the `PostListViewModel` class:

```
@HiltViewModel
class PostListViewModel @Inject constructor(
    ...
) : MviViewModel<PostListModel, UiState<PostListModel>
    , PostListUiAction, PostListUiSingleEvent>() {
    ...
    private fun loadPosts() {
        viewModelScope.launch {
            useCase.execute
            (GetPostsWithUsersWithInteractionUseCase.
            Request)
                .map {
                    converter.convert(it)
                }
                .collect {
                    submitState(it)
                }
        }
    }
}
```

In this snippet, we load the data from
`GetPostsWithUsersWithInteractionUseCase` and collect it and update
`uiStateFlow` through the `submitState` method inherited from the parent class.

13. Implement the `updateInteraction` method in the `PostListViewModel`
class:

```
@HiltViewModel
class PostListViewModel @Inject constructor(
    ...
) : MviViewModel<PostListModel, UiState<PostListModel>
    , PostListUiAction, PostListUiSingleEvent>() {
    ...
    private fun updateInteraction(interaction:
        Interaction) {
        viewModelScope.launch {
```

```
                    updateInteractionUseCase.execute(
                        UpdateInteractionUseCase.Request(
                            interaction.copy(
                                totalClicks = interaction.
                                    totalClicks + 1
                            )
                        )
                    ).collect()
                }
            }
        }
```

In this method, we implement the updateInteraction method, which will submit a new value with an incremented click count using UpdateInteractionUseCase.

14. Modify the PostListScreen method in the PostListScreen file in the list package in the presentation-post module so that it will instead use the submitAction method:

```
@Composable
fun PostListScreen(
    viewModel: PostListViewModel,
    navController: NavController
) {
    LaunchedEffect(Unit) {
        viewModel.submitAction(PostListUiAction.Load)
    }
    viewModel.uiStateFlow.collectAsState().value.let {
        state ->
        CommonScreen(state = state) {
            PostList(it, { postListItem ->
                viewModel.submitAction
                    (PostListUiAction.PostClick
                        (postListItem.id, it.interaction))
            }) { postListItem ->
                viewModel.submitAction
                    (PostListUiAction.UserClick
```

```
                    (postListItem.id, it.interaction))
            }
        }
    }
}
```

Here, we are changing how we interact with `PostListViewModel`. Instead of invoking each separate method for loading and updating the interaction, we instead use the `submitAction` method from `MviViewModel`. In order to load the data, we are using `LaunchedEffect` so that when Jetpack Compose triggers recomposition, the data load won't be retriggered. We are also subscribing to `uiStateFlow` instead of `postListFlow`, which no longer exists.

15. In the same method, subscribe to `singleEventFlow` so that it opens `PostScreen` and `UserScreen` when the appropriate events are received:

```
@Composable
fun PostListScreen(
    viewModel: PostListViewModel,
    navController: NavController
) {
    ...
    LaunchedEffect(Unit) {
        viewModel.singleEventFlow.collectLatest {
            when (it) {
                is PostListUiSingleEvent.
                    OpenPostScreen -> {     navController.
    navigate
                    (it.navRoute)
                }
                is PostListUiSingleEvent.
                    OpenUserScreen -> {
                    navController.navigate
                    (it.navRoute)
                }
            }
        }
    }
}
```

In this snippet, we will need to monitor the events from `singleEventFlow` and then check the events emitted and open the appropriate screen.

16. In the `single` package of the `presentation-post` module, create the `PostUiAction` class and its subclass:

```
sealed class PostUiAction : UiAction {
    data class Load(val postId: Long) : PostUiAction()
}
```

17. In the same package, modify `PostViewModel` so that it extends `MviViewModel`:

```
@HiltViewModel
class PostViewModel @Inject constructor(
    private val postUseCase: GetPostUseCase,
    private val postConverter: PostConverter
) : MviViewModel<PostModel, UiState<PostModel>,
PostUiAction, UiSingleEvent>() {
}
```

Here, we are using the newly created `PostUiAction`, but because we have no one-off events to subscribe to, we will use the `UiSingleEvent` interface.

18. In the same class, implement the `initState` and `handleAction` methods:

```
@HiltViewModel
class PostViewModel @Inject constructor(
    ...
) : MviViewModel<PostModel, UiState<PostModel>,
PostUiAction, UiSingleEvent>() {

    override fun initState(): UiState<PostModel> =
        UiState.Loading

    override fun handleAction(action: PostUiAction) {
        when (action) {
            is PostUiAction.Load -> {
                loadPost(action.postId)
            }
        }
    }
}
```

```
    private fun loadPost(postId: Long) {
        viewModelScope.launch {
            postUseCase.execute
                (GetPostUseCase.Request(postId))
                .map {
                    postConverter.convert(it)
                }
                .collect {
                    submitState(it)
                }
        }
    }
}
```

Here, we are implementing the initState method and returning the UiState. Loading value and the handleAction method. For handleAction, we only have the action to load the data, which will use GetPostUseCase to retrieve the post data and then update uiStateFlow through the submitState method.

19. Modify the PostScreen method from the PostScreen file in the single package in the presentation-post module so that it instead uses the Load action:

```
@Composable
fun PostScreen(
    viewModel: PostViewModel,
    postInput: PostInput
) {
    viewModel.uiStateFlow.collectAsState().value.let {
        result ->
        CommonScreen(result) { postModel ->
            Post(postModel)
        }
    }
    LaunchedEffect(postInput.postId) {
        viewModel.submitAction(PostUiAction.
            Load(postInput.postId))
```

```
        }
    }
```

In this snippet, we are following the same principle as in `PostListScreen` where we replace the interaction with `PostViewModel` to use the `submitAction` method and use `LaunchedEffect` to isolate the data loading.

20. In the `single` package of the `presentation-user` module, create the `UserUiAction` class and its subclass:

```
sealed class UserUiAction : UiAction {


    data class Load(val userId: Long) : UserUiAction()
}
```

21. In the same package, modify `UserViewModel` so that it extends the `MviViewModel` class:

```
@HiltViewModel
class UserViewModel @Inject constructor(
    private val userUseCase: GetUserUseCase,
    private val converter: UserConverter
) : MviViewModel<UserModel, UiState<UserModel>,
    UserUiAction, UiSingleEvent>() {
}
```

Here, we are using the newly created `UserUiAction`, but because we have no one-off events to subscribe to, we will use the `UiSingleEvent` interface.

22. In the same class, implement the `initState` and `handleAction` methods:

```
@HiltViewModel
class UserViewModel @Inject constructor(
    ...
) : MviViewModel<UserModel, UiState<UserModel>,
    UserUiAction, UiSingleEvent>() {

    override fun initState(): UiState<UserModel> =
        UiState.Loading

    override fun handleAction(action: UserUiAction) {
```

```
            when (action) {
                is UserUiAction.Load -> {
                    loadUser(action.userId)
                }
            }
        }

        private fun loadUser(userId: Long) {
            viewModelScope.launch {
                userUseCase.execute
                    (GetUserUseCase.Request(userId))
                    .map {
                        converter.convert(it)
                    }
                    .collect {
                        submitState(it)
                    }
            }
        }
    }
```

Here, we are following the same principle as for `PostViewModel`, which is to implement the `initState` method to return `UiState.Loading`, then in `handleAction`, we check the type, and for the `Load` action, we load the user information.

23. Modify the `UserScreen` method from the `UserScreen` file in the `single` package in the `presentation-user` module so that it instead uses the `Load` action:

```
@Composable
fun UserScreen(
    viewModel: UserViewModel,
    userInput: UserInput
) {
    viewModel.uiStateFlow.collectAsState().value.let {
        result ->
        CommonScreen(result) { userModel ->
```

```
            User(userModel)
        }
    }
    LaunchedEffect(userInput.userId) {
        viewModel.submitAction(UserUiAction.
            Load(userInput.userId))
    }
}
```

In this snippet, we are following the same principle as in `PostScreen` where we replace the interaction with `UserViewModel` to use the `submitAction` method and use `LaunchedEffect` to isolate the data loading.

If we run the application, we will see the same output as in *Exercise 08.02 – Multi-module data presentation*:

Figure 9.2 – Output of Exercise 09.01

After introducing MVI into the exercise, we can see that we already had the groundwork because of how Jetpack Compose requires states to manage the user interface. This represents one of the reasons we ended up creating the `UiState` class in previous chapters. We have also observed the downsides of the pattern through the addition of boilerplate code and the handling of one-off events, the latter not being limited to MVI. The use of `MviViewModel` shows how we can have the same template across different modules of the presentation layer.

From a Clean Architecture perspective, we can see that the changes we have done in our presentation layer haven't affected the rest of the layers of the application, which is a sign that we are going down the right path.

Summary

In this chapter, we studied the MVI architecture pattern and the benefits it provides to applications using reactive streams of data, by centralizing user and application actions into a unidirectional flow of data.

We then looked at how we can implement this pattern using Kotlin flows and the role it plays when combined with other patterns, such as MVP and MVVM, with a focus on MVVM. We can observe the downsides of the pattern on simple presentations, but its benefits become more visible in applications with complicated user interfaces that take in multiple user inputs, which can change the states of other inputs. In the chapter's exercise, we looked at how we can transition an application with MVVM to MVI and how it fits into Clean Architecture.

In the next chapter, we will take a step back and look at what we have implemented and studied so far. We will see what we can improve and how we can take advantage of the different layers of the application, as well as how we can swap dependencies for the various configurations an application might have.

10
Putting It All Together

In this chapter, we will analyze what we have done so far in the previous chapters and look at different ways we can improve the layers of the application. Later, we will explore the benefit of clean architecture when we integrate instrumented testing into the application, where we will swap the data source dependencies with mock dependencies to ensure the reliability of the tests.

In this chapter, we will cover the following topics:

- Inspecting module dependencies
- Instrumentation testing

By the end of the chapter, you will be able to identify and remove external dependencies in the use case layer of the application to enforce the **Common Closure Principle** (CCP) and know how to create instrumented tests on Android with mock data sources.

Technical requirements

The hardware and software requirements are as follows:

- Android Studio Arctic Fox 2020.3.1 Patch 3

The code files for this chapter can be found here: `https://github.com/PacktPublishing/Clean-Android-Architecture/tree/main/Chapter10`.

Check out the following video to see the Code in Action: `https://bit.ly/3sLr0HS`

Inspecting module dependencies

In this section, we will analyze the dependencies used across the different modules in the application created in the previous chapters.

Following *Exercise 09.01 – Transitioning to MVI* from *Chapter 9, Implementing an MVI Architecture*, we now have a fully functioning application split into separate modules, representing different layers. We can analyze the relationship between the different modules by looking at the `dependencies` block in the `build.gradle` file in each module and focusing in particular on the `implementation(project(path: "{module}"))` lines. If we were to draw a diagram, it would look like the following:

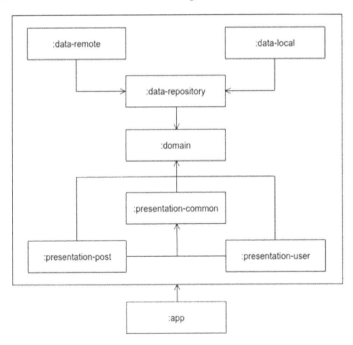

Figure 10.1 – A module dependency diagram for exercise 09.01

In the preceding figure, we can see that the :domain module, which is part of the domain layer, is at the center, with the modules from the other layers having a dependency toward it. The :app module is responsible for assembling all of the dependencies, and this means that it will have a dependency on all the other modules. This means that we are in a good clean architecture position because we want the entities and use cases to have minimal dependencies on other components. If we continue analyzing the build.gradle files for each module and include the external dependencies as well, we will see additional dependencies on external libraries for each module:

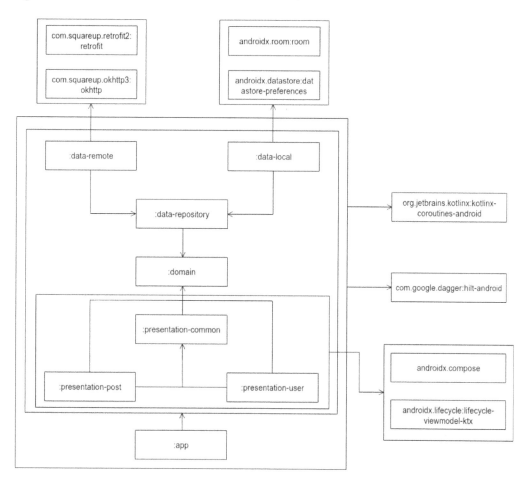

Figure 10.2 – A module dependency diagram with external dependencies for exercise 09.01

In *Figure 10.2*, we can see some of the relevant external dependencies our modules use. `:data-remote` uses dependencies toward Retrofit and OkHttp for networking, the `:data-local` module has dependencies toward Room and DataStore, while the presentation layer modules depend on things such as Compose, ViewModel, and the Android framework. Dependencies that were used used across the entire project were coroutines, flows, and Hilt.

Having dependencies on Hilt and coroutines might pose a problem for the `:domain` and `:data-repository` modules. We want these two modules to be as stable as possible, and having external dependencies will create problems every time we update the versions for those libraries. We decided to use flows because of their threading benefits, the reactive approach, and because they were developed as an extension to the Kotlin framework. They might still pose a problem if we wanted to adapt our use cases for multiple platforms using Kotlin Multiplatform. A solution for this would be to develop a reactive plugin that would abstract the usage of flows and use this abstraction across the different modules. This would allow us to swap different reactive libraries without changing the code inside the module. While this solution would fix the problem, it comes with lots of baggage because we would need to abstract both the streams of data and the operators that the project requires from the flows framework, which would give us more code to maintain.

When it comes to the Hilt dependency, we can remove the references to Hilt from the `:domain` and `:data-repository` modules and move the Hilt modules into `:app`. Another solution would be to create new Gradle modules that would be responsible for providing the necessary dependency. For example, a `:domain-hilt` module could be created, where it would have a `@Module` annotated class that would provide all of the dependencies that the `:domain` module would need to expose. This approach can be used for other modules that we wish to export into applications that use different dependency injection frameworks to avoid the dependency on Hilt in those projects.

Module dependencies will increase as applications develop new features and evolve; this means we should take time and assess the dependencies used in a project. This will help us identify potential issues and whether we can scale an application properly. We should also account for external dependencies and analyze the influence they have over our project. In the following section, we will look at an exercise on how to reduce the dependencies that the domain and repository modules have on Hilt.

Exercise 10.01 – Reduce dependencies

Modify *Exercise 09.01 – transitioning to MVI* in *Chapter 9, Implementing an MVI Architecture*, so that the domain and data-repository modules will no longer depend on Hilt and instead provide the dependencies from those modules inside the app module.

Before completing this exercise, you will need to do the following:

1. Remove Hilt from the domain module.

2. Remove the @Inject annotation from the GetPostsWithUsersWithInteractionUseCase, GetPostUseCase, GetUserUseCase, and UpdateInteractionUseCase classes.

3. Rename the AppModule class UseCaseModule and use @Provides to provide dependencies to the preceding objects.

4. Remove Hilt from the data-repository module and delete the use of the @Inject annotation.

5. Move RepositoryModule from the data-repository module into app and use @Provides to provide the dependencies to PostRepository, UserRepository, and InteractionRepository.

Follow these steps to complete the exercise:

1. In the build.gradle file of the domain module, remove the use of the kapt and Hilt plugins:

```
plugins {
    id 'com.android.library'
    id 'kotlin-android'
}
```

2. In the same file, delete the usages of Hilt from the dependencies block:

```
dependencies {
    implementation coroutines.coroutinesAndroid
    testImplementation test.junit
    testImplementation test.coroutines
    testImplementation test.mockito
}
```

3. Delete the use of `@Inject` from
    `GetPostsWithUsersWithInteractionUseCase`:

```
class GetPostsWithUsersWithInteractionUseCase(
    configuration: Configuration,
    private val postRepository: PostRepository,
    private val userRepository: UserRepository,
    private val interactionRepository:
        InteractionRepository
) : UseCase<GetPostsWithUsersWithInteractionUseCase.
    Request,
        GetPostsWithUsersWithInteractionUseCase.
            Response>(configuration) {
    ...
}
```

4. Delete the use of `@Inject` from `GetPostUseCase`:

```
class GetPostUseCase(
    configuration: Configuration,
    private val postRepository: PostRepository
) : UseCase<GetPostUseCase.Request, GetPostUseCase.
    Response>(configuration) {
    ...
}
```

5. Delete the use of `@Inject` from `GetUserUseCase`:

```
class GetUserUseCase(
    configuration: Configuration,
    private val userRepository: UserRepository
) : UseCase<GetUserUseCase.Request, GetUserUseCase.
    Response>(configuration) {
    ...
}
```

6. Delete the use of `@Inject` from `UpdateInteractionUseCase`:

```
class UpdateInteractionUseCase(
    configuration: Configuration,
```

```
    private val interactionRepository:
        InteractionRepository
) : UseCase<UpdateInteractionUseCase.Request,
    UpdateInteractionUseCase.Response>(configuration) {
    …
}
```

7. In the app module, rename `AppModule` `UseCaseModule`.

8. In the app module in the `UseCaseModule` class, provide a dependency to `GetPostsWithUsersWithInteractionUseCase`:

```
@Module
@InstallIn(SingletonComponent::class)
class UseCaseModule {
    …
    @Provides
    fun
    provideGetPostsWithUsersWithInteractionUseCase(
        configuration: UseCase.Configuration,
        postRepository: PostRepository,
        userRepository: UserRepository,
        interactionRepository: InteractionRepository
    ): GetPostsWithUsersWithInteractionUseCase =
        GetPostsWithUsersWithInteractionUseCase(
        configuration,
        postRepository,
        userRepository,
        interactionRepository
    )
}
```

Here, we need to use `@Provides` because we are no longer in the same module, which means we should treat this as an external dependency, which needs the `@Provides` annotation, similar to how we provided the Room and Retrofit dependencies.

9. In the same class, provide a dependency to `GetPostUseCase`:

```kotlin
@Module
@InstallIn(SingletonComponent::class)
class UseCaseModule {

    ...

    @Provides
    fun provideGetPostUseCase(
        configuration: UseCase.Configuration,
        postRepository: PostRepository
    ): GetPostUseCase = GetPostUseCase(
        configuration,
        postRepository
    )

}
```

In this snippet, we follow the approach of the previous step.

10. In the same class, provide a dependency to `GetUserUseCase`:

```kotlin
@Module
@InstallIn(SingletonComponent::class)
class UseCaseModule {

    ...

    @Provides
    fun provideGetUserUseCase(
        configuration: UseCase.Configuration,
        userRepository: UserRepository
    ): GetUserUseCase = GetUserUseCase(
        configuration,
        userRepository
    )

}
```

In this snippet, we follow the approach of the previous step.

11. In the same class, provide a dependency to `UpdateInteractionUseCase`:

```kotlin
@Module
@InstallIn(SingletonComponent::class)
class UseCaseModule {
```

```
    ...
    @Provides
    fun provideUpdateInteractionUseCase(
        configuration: UseCase.Configuration,
        interactionRepository: InteractionRepository
    ): UpdateInteractionUseCase =
        UpdateInteractionUseCase(
        configuration,
        interactionRepository
    )
}
```

In this snippet, we follow the approach of the previous step.

12. In the build.gradle file of the data-repository module, remove the use of the kapt and Hilt plugins:

```
plugins {
    id 'com.android.library'
    id 'kotlin-android'
}
```

13. In the same file, delete the usages of Hilt from the dependencies block:

```
dependencies {
    implementation(project(path: ":domain"))
    implementation coroutines.coroutinesAndroid
    testImplementation test.junit
    testImplementation test.coroutines
    testImplementation test.mockito
}
```

14. Move the RepositoryModule class from the injection package in the data-repository module into the injection package in the app module and make the class not abstract.

15. Delete the use of @Inject from InteractionRepositoryImpl:

```
class InteractionRepositoryImpl(
    private val interactionDataSource:
LocalInteractionDataSource
```

```
)  : InteractionRepository {
    ...
}
```

16. Delete the use of `@Inject` from `PostRepositoryImpl`:

```
class PostRepositoryImpl(
    private val remotePostDataSource:
        RemotePostDataSource,
    private val localPostDataSource:
        LocalPostDataSource
) : PostRepository {
    ...
}
```

17. Delete the use of `@Inject` from `UserRepositoryImpl`:

```
class UserRepositoryImpl(
    private val remoteUserDataSource:
        RemoteUserDataSource,
    private val localUserDataSource:
        LocalUserDataSource
) : UserRepository {
    ...
}
```

18. In the `RepositoryModule` class, replace the `bindPostRespository` method with a `@Provides` method:

```
@Module
@InstallIn(SingletonComponent::class)
abstract class RepositoryModule {
    @Provides
    fun providePostRepository(
        remotePostDataSource: RemotePostDataSource,
        localPostDataSource: LocalPostDataSource
    ): PostRepository = PostRepositoryImpl(
        remotePostDataSource,
        localPostDataSource
```

```
    )
    ...
}
```

Here, we are no longer able to use the `@Binds` annotation because we removed the `@Inject` annotation from the `PostRepositoryImpl` class, and because it is an external dependency, we will need to use `@Provides`.

19. In the same file, replace the `bindUserRepository` method with a `@Provides` method:

```
@Module
@InstallIn(SingletonComponent::class)
abstract class RepositoryModule {
    ...
    @Provides
    fun provideUserRepository(
        remoteUserDataSource: RemoteUserDataSource,
        localUserDataSource: LocalUserDataSource
    ): UserRepository = UserRepositoryImpl(
        remoteUserDataSource,
        localUserDataSource
    )
    ...
}
```

20. In the same file, replace `bindInteractionRepository` method with a `@Provides` method:

```
@Module
@InstallIn(SingletonComponent::class)
abstract class RepositoryModule {
    ...
    @Provides
    fun provideInteractionRepository(
        interactionDataSource:
            LocalInteractionDataSource
    ): InteractionRepository =
        InteractionRepositoryImpl(
```

```
            interactionDataSource
    )

      ...

}
```

If we run the application, we should see the same output that we got in *Exercise 09.01 – Transitioning to MVI*:

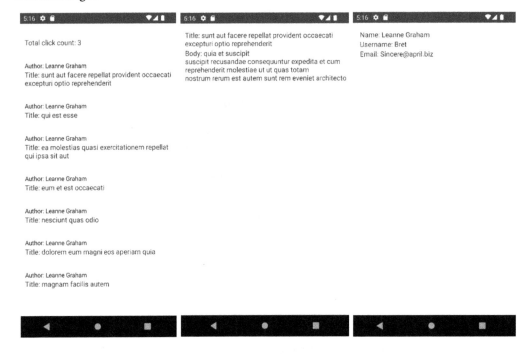

Figure 10.3 – The output of exercise 10.01

The project is now in a state where the `domain` and `data-repository` modules no longer depend on Hilt. This means that all the other modules that depend on these two will be less exposed to potential issues caused by updates to Hilt. It also means that in the future, if we want to change the dependency injection framework used across the application, the `domain` and `data-repository` modules will remain unaffected by the change. In the section that follows, we will look at how we can create instrumentation tests with mock data to test that the modules are well integrated and that the data being passed is processed appropriately.

Instrumentation testing

In this section, we will look at how to perform instrumentation testing for an Android application and how we can take advantage of dependency injection to inject either mock data or add test-related logic without modifying the structure of an application's code.

Instrumentation testing is a set of tests that are run on an Android device or emulator and is represented by the tests written in the `androidTest` directory. Just like other parts of Android development, instrumentation testing evolved across the years to improve the quality of test code and to provide the ability to create better tests and assertions. Initially, testing was done using test classes such as `ActivityTestCase`, `ContentProviderTestCase`, and `ServiceTestCase`, which were mainly used to test individual components of an application in isolation. The addition of the Espresso testing libraries allows us to easily test multiple activities as part of the journey a user would undertake.

In order to add Espresso and the associated libraries into a project, the following will need to be added to any module's `build.gradle` file:

```
dependencies {
    ...
    androidTestImplementation "androidx.test:core:1.4.0"
    androidTestImplementation "androidx.test:runner:1.4.0 "
    androidTestImplementation "androidx.test:rules:1.4.0 "
    androidTestImplementation
        "androidx.test.ext:junit:1.1.3 "
    androidTestImplementation
        "androidx.test.espresso:espresso-core:3.4.0 "
    androidTestImplementation "androidx.test.espresso.
        idling:idling-concurrent:3.4.0 "
}
```

The following is an example of a test written using Espresso:

```
@Test
fun myTest(){
    ActivityScenario.launch(MainActivity::class.java).
        moveToState(Lifecycle.State.RESUMED)
    onView(withId(R.id.my_id))
        .perform(click())
        .check(isDisplayed())
}
```

In the preceding example, we use the `ActivityScenario` launch method to start `MainActivity` and transition `Activity` to the `RESUMED` state. We then use `onView`, which requires `ViewMatcher`, and `withId` looks up `View` by its ID and returns `ViewMatcher` holding that information. We then have the option to use `perform`, which requires `ViewAction`. This is for when we want to interact with certain views. We can also perform `ViewAssertion` using the `check` method. In this case, we are checking whether a view is displayed.

Another useful addition to help with testing is the orchestrator. The orchestrator is useful when we want to delete the data generated by the tests that might be kept in memory or persisted on the device and that in turn might impact other tests and cause them to malfunction. What the orchestrator does is uninstall the application before each executed test so that every test will be on a freshly installed app. In order to add the orchestrator to the application, you will need to add it to the module's `build.gradle` file:

```
android {
    ...
    defaultConfig {
        ...
        testInstrumentationRunnerArguments
            clearPackageData: 'true'
        testOptions {
            execution 'ANDROIDX_TEST_ORCHESTRATOR'
        }
        ...
    }
}
```

This will add the orchestrator configuration into the test execution and pass the instruction to delete the application data after each test. To add the orchestrator dependency into the project, the following is required:

```
dependencies {
    ...

    androidTestUtil "androidx.test:orchestrator: 1.4.1"
}
```

Espresso also comes with many extensions, one of which is the concept of `IdlingResource`. When both local tests (tests that are run on the development machine) and instrumented tests are run, they are run on a dedicated set of threads for testing. The Espresso testing library will monitor the main thread of the application, and when it is idle, it will make the assertions required. If the application uses background threads, Espresso will need a way to be informed by this. We can use `IdlingResource` to indicate to Espresso wait for an action to complete before continuing its execution. An example of `IdlingResource` is `CountingIdlingResource`, which will hold a counter for each operation Espresso will need to wait for. The counter is incremented before each long-running operation and then decremented after the operation is completed. Before each test, `IdlingResource` will need to be registered and then unregistered when the test finishes:

```
class MyClass(private val countingIdlingResource:
    CountingIdlingResource) {

    fun doOperation() {
        countingIdlingResource.increment()
        // Perform long running operation
        countingIdlingResource.decrement()
    }
}
```

In the preceding example, we have `CountingIdlingResource` being incremented at the beginning of the `doOperation` method and decremented after the long operation we intend to perform. To register and unregister `IdlingResource`, we can perform the following:

```
    lateinit var countingIdlingResource :
CountingIdlingResource
```

```
@Before
fun setUp() {
    IdlingRegistry.getInstance().register
        (countingIdlingResource)
}

@After
fun tearDown() {
    IdlingRegistry.getInstance().
        unregister(countingIdlingResource)
}
```

In this example, we register `IdlingResource` in the `setUp` method, which is called before each test because of the `@Before` annotation, and unregister it in the `tearDown` method, which is called after each test because of the `@After` annotation.

Because `IdlingResource` is a part of Espresso but needs to be used when operations inside the application's code are executed, we want to avoid using `IdlingResource` alongside that code. A solution for this is to decorate the class that contains the operation and then use dependency injection to inject the decorated dependency into the test. To decorate the code, we will need to have an abstraction for the operation. An example of this is as follows:

```
interface MyInterface {

    fun doOperation()
}

class MyClass : MyInterface {

    override fun doOperation() {
        // Implement long running operation
    }
}
```

In the preceding example, we have created an interface that defines the doOperation method, and then we implement the interface with the long-running operation into a class. We can now create a class that will belong to the androidTest folder, which will decorate the current implementation of the class:

```
class MyDecoratedClass(
    private val myInterface: MyInterface,
    private val countingIdlingResource:
        CountingIdlingResource
) : MyInterface {

    override fun doOperation() {
        countingIdlingResource.increment()
        myInterface.doOperation()
        countingIdlingResource.decrement()
    }
}
```

Here, we have another implementation of MyInterface, which will hold a reference to the abstraction and CountingIdlingResource. When doOperation is called, we will increment IdlingResource, call the operation, and then, when it's done, decrement IdlingResource.

If we want to inject the new dependency into the test, we will need first to define a new class that extends Application, which will hold the dependency graph containing the test dependencies. If we are using Hilt, it already provides such a class in the form of HiltTestApplication. If we want to integrate Hilt into the instrumented tests, we will need the following dependencies to be added to the module's build.gradle file:

```
dependencies {
    androidTestImplementation "com.google.dagger:hilt-
        android-testing:2.40.5"
    kaptAndroidTest "com.google.dagger:hilt-android-
        compiler: 2.40.5"
}
```

To provide the `HiltTestApplication` class to the test, we will need to change the instrumented test runner. An example of a new test runner will look like the following:

```kotlin
class MyTestRunner : AndroidJUnitRunner() {

    override fun newApplication(cl: ClassLoader?, name:
        String?, context: Context?): Application {
        return super.newApplication(cl,
            HiltTestApplication::class.java.name, context)
    }
}
```

In this example, we are extending from `AndroidJUnitRunner`, and in the `newApplication` method, we invoke the `super` method, and we pass `HiltTestApplication` as the name. This means that when the test is executed, `HiltTestApplication` will be used instead of the `Application` class we defined in our main code. We will now need to change the configuration in the module's `build.gradle` file to use the preceding runner:

```gradle
android {

    ...

    defaultConfig {

        ...

        testInstrumentationRunner "com.test.MyTestRunner"

        ...

    }
}
```

This allows the instrumented test to use the runner we have created. Let's now assume that we have the following module, which will provide the initial dependency:

```kotlin
@Module
@InstallIn(SingletonComponent::class)
abstract class MyModule {

    @Binds
    abstract fun bindMyClass(myClass: MyClass): MyInterface
}
```

Here, we are using a simple binding to connect an implementation to the abstraction. In the `androidTest` folder, we can create a new module in which we replace this instance with the decorated one:

```
@Module
@TestInstallIn(
    components = [SingletonComponent::class],
    replaces = [MyModule::class]
)
class MyDecoratedModule {

    @Provides
    fun provideIdlingResource() =
        CountingIdlingResource("my-idling-resource")

    @Provides
    fun provideMyDecoratedClass(countingIdlingResource:
        CountingIdlingResource) =
        MyDecoratedClass(MyClass(), countingIdlingResource)
}
```

In this example, we use the `@TestInstallIn` annotation, which will make the dependencies in this module live as long as the test application and replace the dependencies in the previous module. We can then provide dependencies for `IdlingResource` and `MyDecoratedClass`, which will wrap `MyClass` and use `IdlingResource`. If we want these changes to take effect in the tests, we will need the following changes:

```
@HiltAndroidTest
class MyActivityTest {

    @get:Rule(order = 0)
    var hiltAndroidRule = HiltAndroidRule(this)

    @Inject
    lateinit var idlingResource: CountingIdlingResources

    @Before
```

```
    fun setUp() {
        hiltAndroidRule.inject()
        IdlingRegistry.getInstance().register
            (idlingResource)
    }

    @After
    fun tearDown() {
        IdlingRegistry.getInstance().unregister
            (idlingResource)
    }
}
```

In this example, we have used the `@HiltAndroidTest` annotation because we want to inject `CountingIdlingResources` into the test. We then used `HiltAndroidTestRule` to perform the injection. We also gave it the highest priority in terms of the order of execution for test rules. Finally, we were able to register and unregister `CountingIdlingResources` for each test in the class.

Jetpack Compose comes with its own testing libraries, which require the following configuration to the module's `build.gradle` file:

```
dependencies {
    androidTestImplementation "androidx.compose.ui:ui-test-
        junit4:1.0.5"
    debugImplementation "androidx.compose.ui:ui-test-
        manifest:1.0.5"
}
```

To write tests for Jetpack Compose components, we will need to define a Compose test rule using `createComposeRule` when we want to test individual composable methods, or `createAndroidComposeRule` if we want to test the Compose content of an entire activity. An example would look like the following:

```
class MyTest {

    @get:Rule
    var composeTestRule = createAndroidComposeRule
```

```
        (MyActivity::class.java)
}
```

In the preceding example, we have defined a test rule that will be responsible for testing the Compose content inside `MyActivity`. If we want the test to interact with the user interface or assert that it displays the correct information, we have the following structure:

```
@Test
fun testDisplayList() {
    composeTestRule.onNode()
        .assertIsDisplayed()
        .performClick()
}
```

In this example we use the `onNode` method to locate a particular element, such as `Text` or `Button`. We then have the `assertIsDisplayed` method, which is used to check whether the node is displayed. Finally, we have the `performClick` method, which will click on the element. Jetpack Compose uses its own `IdlingResource` type, which can be registered in the Compose test rule, similar to the following example:

```
lateinit var idlingResource: IdlingResource

@Before
fun setUp() {
    composeTestRule.registerIdlingResource
        (idlingResource)
}

@After
fun tearDown() {
    composeTestRule.unregisterIdlingResource
        (idlingResource)
}
```

From a clean architecture perspective, we should strive to make our application's code as testable as possible. This applies to both local tests such as unit tests and instrumented tests. We want to be able to ensure that the tests are reliable; this usually means that we will need to remove the dependency on network calls, which means we will need to provide a way to inject mock data into the application without modifying the application's code. We also need to be able to either inject `IdlingResources` into the application or use decorated dependencies to verify that the data inserted by the user is the correct data received in the data layer. This also involves the ability to decorate these dependencies to add extra logic without modifying the application's code. In the following section, we will look at an exercise in which we will inject various dependencies containing testing logic into the application and assess the difficulty it takes to introduce them.

Exercise 10.02 – Instrumented testing

Add one instrumented test to *Exercise 10.01 – Reduce dependencies*, which will assert that the following data is displayed onscreen:

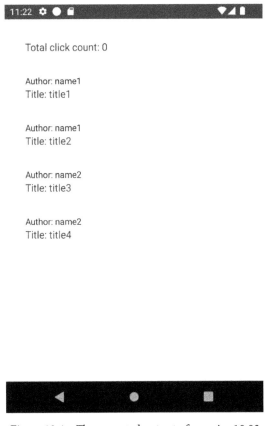

Figure 10.4 – The expected output of exercise 10.02

To achieve this, you will need to create a new implementation of
`RemotePostDataSource`, which will return a list of four posts; two posts will belong
to one user and the other two will belong to another user. The same thing will need
to be done for `RemoteUserDataSource`, which will return the two users. These
implementations will need to be injected into the test. To ensure that the test will wait
for the background work to complete, you will need to decorate each repository with
`IdlingResource`, which will also need to be injected into the test.

Before completing this exercise, you will need to do the following:

1. Integrate the testing libraries into the app module.

2. Create `PostAppTestRunner`, which will be used to provide
   `HiltTestApplication` to the Android instrumentation test runner.

3. Create a `ComposeCountingIdlingResource` class, which will wrap
   an Espresso `CountingIndlingResource` and implement the Compose
   `IdlingResource`.

4. Create `MockRemotePostDataSource` and `MockRemoteUserDataSource`,
   which will be responsible for returning the users and posts in presented in *Figure 10.4.*

5. Create `IdlingInteractionRepository`, `IdlingUserRepository`, and
   `IdlingPostRespository`, which will decorate `InteractionRepository`,
   `UserRepository`, and `PostRepository`, and use the
   `ComposeCountingIdlingResource`, which will be incremented when new
   data is loaded and decremented when data loading is done.

6. Create `IdlingRepositoryModule` and `MockRemoteDataSourceModule`,
   which will replace `RepositoryModule` and `RemoteDataSourceModule`
   respectively in the tests.

7. Create `MainActivityTest`, which will have one test, and use
   `createAndroidComposeRule` to assert that the list of mock data is displayed.

Follow these steps to complete the exercise:

1. In the top-level `build.gradle` file, add the following library versions:

```
buildscript {
    ext {
        ...
        versions = [
            ...
```

```
            androidTestCore       : "1.4.0",
            androidTestJunit      : "1.1.3",
            orchestrator          : "1.4.1"
        ]
        ...
}
```

2. In the same file, make sure that the following `androidTest` dependencies are added:

```
buildscript {
    ext {
        ...
        androidTest = [
                junit                : "androidx.test.ext
                    :junit:${versions.espressoJunit}",
                espressoCore         : "androidx.test.
                    espresso:espresso-core:${versions.
                        espressoCore}",
                idlingResource       : "androidx.test.
                    espresso:espresso-idling-resource
                        :${versions.espressoCore}",
                composeUiTestJunit: "androidx.compose.
                    ui:ui-test-junit4:$
                        {versions.compose}",
                composeManifest      : "androidx.compose
                    .ui:ui-test-manifest:$
                        {versions.compose}",
                hilt                 : "com.google.
                    dagger:hilt-android-testing:$
                        {versions.hilt}",
                hiltCompiler         : "com.google.
                    dagger:hilt-android-compiler:$
                        {versions.hilt}",
                core                 : "androidx.test:
                    core:${versions.androidTestCore}",
                runner               : "androidx.test:
```

```
                    runner:$
                        {versions.androidTestCore}",
                rules             : "androidx.test:
                    rules:${versions.androidTestCore}",
                orchestrator       : "androidx.test:
                    orchestrator:$
                        {versions.orchestrator}"
        ]
    }
    ...
}
```

Here, we are defining the mappings for all the testing libraries we will be using so that they will be available across multiple modules.

3. In the `build.gradle` file of the app module, add the required test dependencies:

```
dependencies{
    ...
    androidTestImplementation androidTest.junit
    androidTestImplementation androidTest.espressoCore
    androidTestImplementation
        androidTest.idlingResource
    androidTestImplementation androidTest.core
    androidTestImplementation androidTest.rules
    androidTestImplementation androidTest.runner
    androidTestImplementation androidTest.hilt
    kaptAndroidTest androidTest.hiltCompiler
    androidTestImplementation
        androidTest.composeUiTestJunit
    debugImplementation androidTest.composeManifest
    androidTestUtil androidTest.orchestrator
}
```

4. In the `androidTest` folder of the app module, create the `PostAppTestRunner` class inside the `java/{package-name}` folder:

```
class PostAppTestRunner : AndroidJUnitRunner() {
```

```
        override fun newApplication(cl: ClassLoader?,
            name: String?, context: Context?): Application {
            return super.newApplication(cl,
                HiltTestApplication::class.java.name,
                    context)
        }
    }
```

5. In the `build.gradle` file of the app module, set the following test configuration. Make sure to replace `{package-name}` with the package that the `PostAppTestRunner` is in:

```
android {
    ...
    defaultConfig {
        ...
        testInstrumentationRunner "{package-name}.
            PostAppTestRunner"
        testInstrumentationRunnerArguments
            clearPackageData: 'true'
        testOptions {
            execution 'ANDROIDX_TEST_ORCHESTRATOR'
        }
    }
}
```

6. In the `androidTest` folder of the app module, create the following packages inside the `java/{package-name}` folder – `idling`, `injection`, `remote`, `repository`, and `test`.

7. Inside the `idling` package, create a new class called `ComposeCountingIdlingResource`:

```
class ComposeCountingIdlingResource(name: String) :
    IdlingResource {

    private val countingIdlingResource =
        CountingIdlingResource(name)


    override val isIdleNow: Boolean
```

```
        get() = countingIdlingResource.isIdleNow

    fun increment() = countingIdlingResource.
        increment()

    fun decrement() = countingIdlingResource.
        decrement()
}
```

Here, we have used the `CountingIdlingResource` class from Espresso to perform the logic for incrementing, decrementing, and providing its current idling state through the `isIdleNow` method, which is used by Jetpack Compose.

8. In the same package, create a file called `IdlingUtils` with the following method:

```
fun <T> Flow<T>.attachIdling(
    countingIdlingResource:
        ComposeCountingIdlingResource
): Flow<T> {
    return onStart {
        countingIdlingResource.increment()
    }.onEach {
        countingIdlingResource.decrement()
    }
}
```

This is an extension function that we can use to increment `IdlingResource` before `Flow` is collected and decrement it when the first value of `Flow` is being emitted.

9. In the `repository` package, create a class called `IdlingInteractionRepository`:

```
class IdlingInteractionRepository(
    private val interactionRepository:
InteractionRepository,
    private val countingIdlingResource:
ComposeCountingIdlingResource
) : InteractionRepository {

    override fun getInteraction(): Flow<Interaction> {
```

```
        return interactionRepository.getInteraction()
            .attachIdling(countingIdlingResource)
    }


    override fun saveInteraction(interaction:
        Interaction): Flow<Interaction> {
        return interactionRepository.
            saveInteraction(interaction)
            .attachIdling(countingIdlingResource)
    }
}
```

This class has a reference to the `ComposeCountingIdlingResource` object and the `attachIdling` method created previously to increment when the data is loaded or saved and to decrement when it's done performing these operations.

10. In the same package, create a class called `IdlingPostRepository`:

```
class IdlingPostRepository(
    private val postRepository: PostRepository,
    private val countingIdlingResource:
        ComposeCountingIdlingResource
) : PostRepository {
    override fun getPosts(): Flow<List<Post>> =
        postRepository.getPosts().attachIdling
            (countingIdlingResource)


    override fun getPost(id: Long): Flow<Post> =
        postRepository.getPost(id).
            attachIdling(countingIdlingResource)
}
```

In this snippet, we follow the approach of the previous step.

11. In the same package, create a class called `IdlingUserRepository`:

```
class IdlingUserRepository(
    private val userRepository: UserRepository,
    private val countingIdlingResource:
```

```
            ComposeCountingIdlingResource
    )  :  UserRepository {
        override fun getUsers(): Flow<List<User>> =
            userRepository.getUsers()
                .attachIdling(countingIdlingResource)

        override fun getUser(id: Long): Flow<User> =
            userRepository.getUser(id)
                .attachIdling(countingIdlingResource)
    }
```

In this snippet, we follow the approach of the previous step.

12. In the injection package, create the IdlingRepositoryModule class:

```
@Module
@TestInstallIn(
    components = [SingletonComponent::class],
    replaces = [RepositoryModule::class]
)
class IdlingRepositoryModule {
}
```

13. In the IdlingRepositoryModule class, provide a dependency for ComposeCountingIdlingResource, which will be a single instance across all the repositories:

```
@Module
@TestInstallIn(
    components = [SingletonComponent::class],
    replaces = [RepositoryModule::class]
)
class IdlingRepositoryModule {

    @Singleton
    @Provides
    fun provideIdlingResource():
        ComposeCountingIdlingResource =
```

```
            ComposeCountingIdlingResource
                ("repository-idling")
    }
```

In this snippet, we are providing a single instance of
`ComposeCountingIdlingResource` so that when multiple repositories load
data at the same time, the same counter will be used for all of them.

14. In the same file, provide a dependency for `IdlingPostRepository`:

```
@Module
@TestInstallIn(
    components = [SingletonComponent::class],
    replaces = [RepositoryModule::class]
)
class IdlingRepositoryModule {
    ...
    @Provides
    fun providePostRepository(
        remotePostDataSource: RemotePostDataSource,
        localPostDataSource: LocalPostDataSource,
        countingIdlingResource:
            ComposeCountingIdlingResource
    ): PostRepository = IdlingPostRepository(
        PostRepositoryImpl(
            remotePostDataSource,
            localPostDataSource
        ),
        countingIdlingResource
    )
}
```

In this snippet, we are providing an instance of the `IdlingPostRepository`,
which will wrap an instance of `PostRepositoryImpl` and have a reference to the
`ComposeCountingIdlingResource` instance defined previously.

15. In the same file, provide a dependency for `IdlingUserRepository`:

```
@Module
@TestInstallIn(
```

```
        components = [SingletonComponent::class],
        replaces = [RepositoryModule::class]
    )
    class IdlingRepositoryModule {
        ...

        @Provides
        fun provideUserRepository(
            remoteUserDataSource: RemoteUserDataSource,
            localUserDataSource: LocalUserDataSource,
            countingIdlingResource:
                ComposeCountingIdlingResource
        ): UserRepository = IdlingUserRepository(
            UserRepositoryImpl(
                remoteUserDataSource,
                localUserDataSource
            ),
            countingIdlingResource
        )
    }
```

In this snippet, we are providing an instance of `IdlingUserRepository`, which will wrap an instance of `UserRepositoryImpl` and have a reference to the `ComposeCountingIdlingResource` instance defined previously.

16. In the same file, provide a dependency for `IdlingInteractionRepository`:

```
@Module
@TestInstallIn(
    components = [SingletonComponent::class],
    replaces = [RepositoryModule::class]
)
class IdlingRepositoryModule {
    ...

    @Provides
    fun provideInteractionRepository(
        interactionDataSource:
            LocalInteractionDataSource,
        countingIdlingResource:
```

```
                    ComposeCountingIdlingResource
        ): InteractionRepository =
            IdlingInteractionRepository(
            InteractionRepositoryImpl(
                interactionDataSource
            ),
            countingIdlingResource
        )
    }
```

In this snippet, we are providing an instance of
`IdlingInteractionRepository`, which will wrap an instance
of `InteractionRepositoryImpl` and have a reference to the
`ComposeCountingIdlingResource` instance defined previously.

17. In the `remote` package, create a class called `MockRemoteUserDataSource` and create a list of `User` objects representing the test data:

```
class MockRemoteUserDataSource @Inject constructor() :
    RemoteUserDataSource {

    private val users = listOf(
        User(
            id = 1L,
            name = "name1",
            username = "username1",
            email = "email1"
        ),
        User(
            id = 2L,
            name = "name2",
            username = "username2",
            email = "email2"
        )
    )


    override fun getUsers(): Flow<List<User>> = flowOf
```

```
        (users)

    override fun getUser(id: Long): Flow
        flowOf(users[0])
}
```

Here, we have created a list in which we return two users and put it into `Flow` for the `getUsers` method.

18. In the same package, create a class called `MockRemotePostDataSource` and create a list of `Post` objects representing the test data:

```
class MockRemotePostDataSource @Inject constructor() :
    RemotePostDataSource {

    private val posts = listOf(
        Post(
            id = 1L,
            userId = 1L,
            title = "title1",
            body = "body1"
        ),
        Post(
            id = 2L,
            userId = 1L,
            title = "title2",
            body = "body2"
        ),
        Post(
            id = 3L,
            userId = 2L,
            title = "title3",
            body = "body3"
        ),
        Post(
            id = 4L,
            userId = 2L,
            title = "title4",
```

```
                    body = "body4"
            )
        )

    override fun getPosts(): Flow<List<Post>> =
        flowOf(posts)

    override fun getPost(id: Long): Flow<Post> =
        flowOf(posts[0])
}
```

Similar to what we did with the users, we create a list of posts and connect the first two posts to the first user and the last two posts to the second user.

19. In the `injection` package, create a class called `MockRemoteDataSourceModule`, which will be responsible for binding the previous two implementations to the abstractions:

```
@Module
@TestInstallIn(
    components = [SingletonComponent::class],
    replaces = [RemoteDataSourceModule::class]
)
abstract class MockRemoteDataSourceModule {

    @Binds
    abstract fun bindPostDataSource(
        postDataSourceImpl: MockRemotePostDataSource):
            RemotePostDataSource

    @Binds
    abstract fun bindUserDataSource(userDataSourceImpl
        : MockRemoteUserDataSource):
            RemoteUserDataSource
}
```

20. In the `test` package, create a class called `MainActivityTest`:

```kotlin
@HiltAndroidTest
class MainActivityTest {

    @get:Rule(order = 0)
    var hiltAndroidRule = HiltAndroidRule(this)

    @get:Rule(order = 1)
    var composeTestRule = createAndroidComposeRule
        (MainActivity::class.java)

    @Inject
    lateinit var idlingResource:
        ComposeCountingIdlingResource

    @Before
    fun setUp() {
        hiltAndroidRule.inject()
        composeTestRule.
            registerIdlingResource(idlingResource)
    }

    @After
    fun tearDown() {
        composeTestRule.unregisterIdlingResource
            (idlingResource)
    }
}
```

Here, we are initializing our test rules, which are for Hilt and Compose, in that exact order. Then, we inject `ComposeCountingIdlingResource` into the test class so that we can register it into the Compose test rule.

21. In the `MainActivityTest` class, add a test that will assert that the required data is displayed on the screen:

```
@HiltAndroidTest
class MainActivityTest {
    ...
    @Test
    fun testDisplayList() {
        composeTestRule.onNodeWithText("Total click
            count: 0")
            .assertIsDisplayed()
        composeTestRule.onAllNodesWithText("Author:
            name1")
            .assertCountEquals(2)
        composeTestRule.onAllNodesWithText("Author:
            name2")
            .assertCountEquals(2)
        composeTestRule.onNodeWithText("Title:
            title1")
            .assertIsDisplayed()
        composeTestRule.onNodeWithText("Title:
            title2")
            .assertIsDisplayed()
        composeTestRule.onNodeWithText("Title:
            title3")
            .assertIsDisplayed()
        composeTestRule.onNodeWithText("Title:
            title4")
            .assertIsDisplayed()
    }
}
```

Here, we have added a test that asserts that the header text is displayed, the two users are displayed for each of their posts, and that each post is displayed.

If we run the test, we should see the following output:

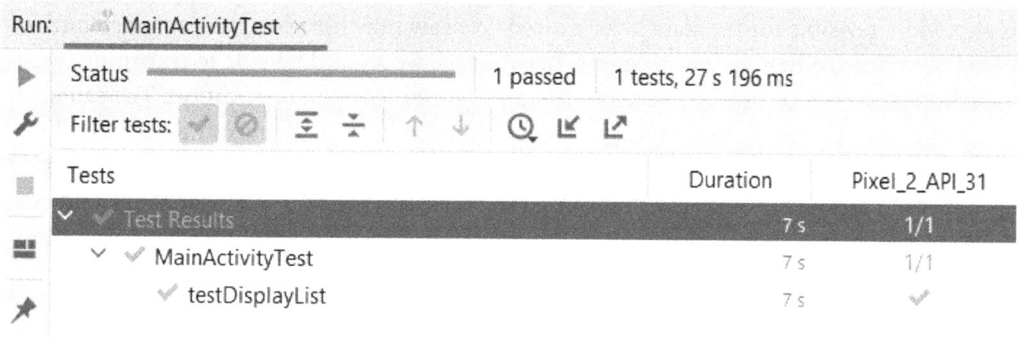

Figure 10.5 – The test output for exercise 10.02

As part of this exercise, we were able to provide mock data to the application without changing any of its existing code by changing the remote data sources and building upon existing functionality, by adding `IdlingResources` to our repositories. Both techniques were possible using dependency injection, and because of abstractions, we introduced different layers of the application when we performed dependency inversion. This makes the application's code testable and provides us with the opportunity to test different scenarios and create various types of tests to ensure the integration of different components.

Summary

In this chapter, we analyzed the exercises we've done in previous chapters and found potential issues with the dependencies that the modules of the application have. We looked at potential solutions for these problems. Then, we looked at a practical application of clean architecture, which is the implementation of instrumented tests, and how we can change the data sources of an application to ensure testing reliability. We looked at how we can implement instrumented tests using Jetpack Compose and Hilt to provide dependency injection, and then we applied them in an exercise in which we changed the dependencies for the tests. This serves as just one example of the benefits of clean architecture. Other benefits will come in situations where multiple flavors are used to publish similar applications and want to inject different implementations or configurations for each application we want to build. Another benefit comes when dealing with multiple platforms (such as Android and iOS), where we can define entities, use cases and repositories agnostically of the platforms using cross platform frameworks and then inject the implementations for retrieving and persisting the data for each platform.

In *Chapter 9, Implementing an MVI Architecture*, we showed how we can change an application's presentation layer without impacting other layers. In a clean application, this should be possible for the data layer as well. We saw how libraries have changed and evolved over time. When networking libraries change, we should be able to transition to new libraries without causing issues in the other modules of an application. The same principle can be applied to local storage. We should be able to change from Room to other ways of persisting data locally. A good rule of thumb for how modules should be created is to view each module as a library that can be released and imagine yourself as the end user. You should have now a good idea of how clean architecture is supposed to work, the problems it is trying to solve, and how you can apply it to an Android application.

Index

`Packt.com`

Subscribe to our online digital library for full access to over 7,000 books and videos, as well as industry leading tools to help you plan your personal development and advance your career. For more information, please visit our website.

Why subscribe?

- Spend less time learning and more time coding with practical eBooks and Videos from over 4,000 industry professionals

- Improve your learning with Skill Plans built especially for you

- Get a free eBook or video every month

- Fully searchable for easy access to vital information

- Copy and paste, print, and bookmark content

Did you know that Packt offers eBook versions of every book published, with PDF and ePub files available? You can upgrade to the eBook version at `packt.com` and as a print book customer, you are entitled to a discount on the eBook copy. Get in touch with us at `customercare@packtpub.com` for more details.

At `www.packt.com`, you can also read a collection of free technical articles, sign up for a range of free newsletters, and receive exclusive discounts and offers on Packt books and eBooks.

Other Books You May Enjoy

If you enjoyed this book, you may be interested in these other books by Packt:

Android UI Development with Jetpack Compose.

Thomas Künneth

ISBN: 978-1-80181-216-0

- Gain a solid understanding of the core concepts of Jetpack Compose
- Develop beautiful, neat, and immersive UI elements that are user friendly, reliable, and performant
- Build a complete app using Jetpack Compose
- Add Jetpack Compose to your existing Android applications
- Test and debug apps that use Jetpack Compose

Kickstart Modern Android Development with Jetpack and Kotlin

Catalin Ghita

ISBN: 978-1-80181-107-1

- Integrate popular Jetpack libraries such as Compose, ViewModel, Hilt, and Navigation into real Android apps with Kotlin

- Apply modern app architecture concepts such as MVVM, dependency injection, and clean architecture

- Explore Android libraries such as Retrofit, Coroutines, and Flow

- Integrate Compose with the rest of the Jetpack libraries or other popular Android libraries

- Work with other Jetpack libraries such as Paging and Roo

Packt is searching for authors like you

If you're interested in becoming an author for Packt, please visit authors. packtpub.com and apply today. We have worked with thousands of developers and tech professionals, just like you, to help them share their insight with the global tech community. You can make a general application, apply for a specific hot topic that we are recruiting an author for, or submit your own idea.

Hi!

I am Alexandru Dumbravan, author of *Clean Android Architecture: Take a layered approach to writing a clean, testable, and decoupled Android applications.* I really hope you enjoyed reading this book and found it useful for building Android applications that are easier to test, scale and maintain, and gained an understanding on what roles some of the most popular libraries and frameworks play in achieving this.

It would really help us (and other potential readers!) if you could leave a review on Amazon sharing your thoughts on how to build a clean Android application.

Go to the link below or scan the QR code to leave your review:

`https://packt.link/r/180323458X`

Your review will help us to understand what's worked well in this book, and what could be improved upon for future editions, so it really is appreciated.

Best wishes,

Alexandru Dumbravan